The Threat is Real

Scientific evidence indicates that drinking-water standards do not protect the nation's public health. The only viable alternative is to use currently available advanced water-treatment technologies to provide the purest water possible to each American consumer. The substitution of a "technology-based program" for the current "standards-based policy" will require POLITICAL ACTION.

"The integrity of the environment is not just another issue to be used in political games for popularity, votes, or attention. And the time has long since come to take more political risks—and endure much more political criticism—by proposing tougher, more effective solutions and fighting hard for their enactment."

— Al Gore, *Earth In the Balance*, 1992

AMERICA'S THREATENED DRINKING WATER

Hazards and Solutions

Patrick J. Sullivan, Ph.D.

Franklin J. Agardy, Ph.D.

James J. J. Clark, Ph.D.

Human Habitat and Environmental Health Series

Disclaimer: The discussion of point-of-use or point-of-enter water treatment systems is intended for informational purposes only. The reader should not act or rely on information in this publication without first seeking the advice of a professional water treatment system contractor or treatment unit provider. The authors have endeavored to provide sound scientific commentary relative to point-of-use and point-of-entry drinking water treatment systems. Because cost and risk tolerance may vary among individuals and entities, the available technologies change over time, and individual treatment system performance is highly dependent upon the specific treatment units, maintenance and system configuration, please consult your professional water treatment system contractor or treatment unit provider for the performance, technology, maintenance and testing requirements needed to meet your drinking water treatment objectives.

National Library of Canada Cataloguing in Publication

Sullivan, Patrick J., Ph.D
        America's threatened drinking water : hazards and solutions / Patrick J. Sullivan, Franklin J. Agardy, James J.J. Clark.

Includes bibliographical references and index.
ISBN 1-55369-616-6

        1. Drinking water--United States. 2. Drinking water--Contamination--United States. 3. Water quality management--United
States. I. Agardy, Franklin J. II. Clark, James J. J. III. Title.

RA592.A1S94 2002   363.6'1'0973   C2002-902580-X

TRAFFORD

This book was published *on-demand* in cooperation with Trafford Publishing.
On-demand publishing is a unique process and service of making a book available for retail sale to the public taking advantage of on-demand manufacturing and Internet marketing.
On-demand publishing includes promotions, retail sales, manufacturing, order fulfilment, accounting and collecting royalties on behalf of the author.

Suite 6E, 2333 Government St., Victoria, B.C. V8T 4P4, CANADA
Phone   250-383-6864   Toll-free   1-888-232-4444 (Canada & US)
Fax   250-383-6804   E-mail   sales@trafford.com
Web site   www.trafford.com   TRAFFORD PUBLISHING IS A DIVISION OF TRAFFORD HOLDINGS LTD.
Trafford Catalogue #02-0429   www.trafford.com/robots/02-0429.html

10   9   8   7   6   5   4

Foreword

We all agree that drinking water is threatened, but there is general disagreement about the significance of the threat and what to do about it. 20th-century regulatory schemes were developed to protect the public against known and potential threats; founded on the concept that government should specify the ends not the means by which they are achieved. The recent EPA decision to reduce arsenic levels to a standard of 10 micrograms per liter is but one example of a controversial attempt to establish a national end point in the context of widely varying raw water quality and utility circumstances. An arsenic reduction was clearly supported by the science, but the question was how much, and how best to achieve the reduction which would impose large costs on small rural towns.

It would be hard to find a water professional, scientist, engineer, or manager who thinks that the present permit system really works. It doesn't satisfy public expectations, lags behind in our ability to detect threats, the emergence of new technologies, and our understanding of risk. Even the developing world, trying to achieve in the 21st century what we achieved in the last one, sometimes adopts the same complex regulatory scheme that we use. In other parts of the developed world, European and Japanese water systems deliver more or less the same high quality product as in the United States, but they go about it differently with best and appropriate technology given primary consideration.

Is the time-consuming establishment of increasingly stringent individual contaminant standards likely to result in better treatment or source selection? Does the present system allocate resources to assure efficient and affordable improvements in drinking water quality? When risk reduction is measured by the potential health effect on one in one million consumers, based on consumption of a single contaminant when each glass of water may have 100 or more potential contaminants, is the public well served?

The authors have done a first-rate job of describing of the limitations of the present U.S. regulatory scheme and have arrived at the inevitable conclusion that there should be a fundamental change in the way drinking water improvement is achieved. This book is essentially a challenge to creative water experts to find a way to achieve fundamental changes within our complex system of competing interests.

Jerome B. Gilbert, May 7, 2002
Past President of the American Water Works Association
and former Executive Director of the California State
Water Resources Control Board

v

Contents

Appendices

About the Authors

Patrick J. Sullivan received his B.S. in Geochemistry (1974) and a Ph.D. in Soil Chemistry (1978) from the University of California at Riverside. He was a senior environmental analyst at the Jet Propulsion Laboratory, California Institute of Technology, and later taught Environmental and Soil Sciences at Ball State University, leaving the faculty as a tenured associate professor in 1985. He was Manager of Environmental Chemistry at Western Research Institute at the University of Wyoming for three years. Since 1988, he has been an environmental forensic expert. Over the last 25 years, Dr. Sullivan's environmental research has focused on hazardous substances and water chemistry.

Franklin J. Agardy received his B.S. in Civil Engineering from the City College of New York in 1955, an M.S. (1958) and Ph.D. (1963) in Sanitary Engineering from the University of California at Berkeley. He taught Civil and Sanitary Engineering at San Jose State University and left the faculty as a tenured full professor in 1971. He spent 19 years with URS Corporation, retiring in 1988 as President of the Corporation. Dr. Agardy is a past president of the California Water Pollution Control Association, a past Chairman of the California/Nevada Section of the American Water Works Association and the author of a 1972 paper titled, "The Threat from Additions of Chemicals and Biologicals to a Municipal Water Supply."

James J. J. Clark received his B.S. in Biophysical and Biochemical Sciences from the University of Houston (1987), an M.S. degree in Environmental Health Sciences from the University of California Los Angeles School of Public Health (1993), and a Ph.D. degree in Environmental Health Sciences from the University of California Los Angeles School of Public Health (1995). Dr. Clark is currently a senior toxicologist for Komex H2O Sciences, Inc. in Los Angeles, California and is the author of several recent chapters on emerging drinking water contaminant issues including "TBA: Chemical Properties, Production & Use, Fate and Transport, Toxicology, Detection in Groundwater, and Regulatory Standards" in *Fuel Oxygenates* and "Toxicology of Perchlorate" in *Perchlorate in the Environment*.

Preface

Prior to the 1900s, the dominant chemical pollutants were inorganic and coal-derived organic chemicals. As chemical manufacturing blossomed during the mid-1900s, a wide variety of synthetic petroleum-based chemicals were produced and released to the environment. Today, our nation's chemical manufacturers produce approximately 87,000 different chemicals. Current estimates predict the manufacture of some 2,000 new chemical compounds each year. With so many chemicals being produced and used, it is no wonder that a significant portion of the nation's lakes, rivers and groundwater contain a wide range of industrial inorganic and organic compounds, including pesticides and pharmaceuticals. These same water resources serve as the source for community drinking water systems and private groundwater wells.

In 1925, the United States Public Health Service's Primary Drinking Water Standards required that the concentration of three trace metals not exceed certain limits in drinking water. This requirement could be considered to have been as the first step in controlling chemical pollution. By 1962, nine trace metals and inorganic compounds were included in the drinking water standards. Today, in compliance with the 1974 Safe Drinking Water Act and its amendments, the Primary Drinking Water Standards still only limit the concentration of a very small number of inorganic chemicals, industrial organic compounds and organic pesticides. These regulated compounds represent the major sources of pollution from chemical usage between the 1930s and the1980s. Since this time period, the number of industrial chemicals, pesticides and pharmaceuticals that have been produced and are now found in our waters has increased dramatically. Thus, many of the chemical pollutants found in our waters today are not regulated by current standards.

In 1998, the United States Environmental Protection Agency (USEPA) proposed adding 50 chemicals to the current list of Primary Drinking Water Standards. Yet, pharmaceuticals do not even appear on this list of 50. Given the number of chemicals being used in the United States today and the apparent inability of drinking water standards to rapidly adapt to these changes, it is clear that drinking water standards, as currently implemented, are incapable of keeping pace with the unregulated chemicals that are currently found in our drinking water. Until the U.S. Geological Survey published the results of their 2000 river sampling program for toxic chemicals, the general public was not even aware of the presence of this vast array of unregulated organic chemicals. Yet, because the concentrations of these chemicals are very low, the public was told that it was unlikely that these chemicals would have any impact on their health.

Approximately 50 years ago, the National Agricultural Chemicals Association argued that pesticides in low concentrations would have no effect on wildlife and that cancer, which had afflicted mankind for centuries, could not be shown to be caused by pesticides. After decades of research, these opinions have both been proven to be wrong. Today, one of the main reasons the public is told that these new chemicals are unlikely to affect their health is because quantifying the effects of such low levels of chemical contamination is nearly impossible. We believe that the likelihood of adverse health effects is far greater than advertised but that it will take decades of research before this link is proven.

Thus, it may not be in the public's best interest to assume there are no health effects from the consumption of low levels of man-made chemicals. Indeed, a January 2002 report by the Environmental Working Group entitled, "Consider the Source, Farm Runoff, Chlorination Byproducts, and Human Health," concluded that chemical pollutants in drinking water pose a serious health threat to the American Public in general and pregnant women in particular. As a result, this report recommends that a national health tracking system be set up to better quantify the health impacts of drinking water pollutants. This recommendation, however, skirts the real issue, which is that drinking water in many parts of the United States is polluted by both regulated and unregulated chemicals. Thus, the safety or the purity of drinking water is not guaranteed even if the standards are met.

Indeed, when attempting to determine the government's approach toward protecting public health, one is confronted with a bewildering web of regulatory programs and a sea of acronyms. For example, the quality of water delivered to a consumer is dependent upon whether the water comes from a Community Water Systems (CWS), Nontransient-Noncommunity Water Systems (NTNCWS) and Transient-Noncommunity Water Systems (TNCWS), and depending upon the size of the system the chemical monitoring requirements may vary substantially. In addition, some water systems must comply with the enhanced surface water treatment rule (Stage 2 D/DBPR and the LT2ESWTR), the Groundwater Disinfection Rule (GWDR), and the Information Collection Rule (ICR) when attempting to meet "primary" standards. To further complicate the issue the USEPA has established Maximum Contaminant Levels (MCLs) for various chemicals as well as nonenforceable Maximum Contaminant Level Goals (MCLGs). They also recommend a Best Available Technology (BAT) by which water systems can try to meet these MCLs and MCLGs. The artificial complexity and cost of this compliance system alone is reason enough for why a new system is needed.

Given this situation, the only reasonable recommendation is to treat drinking water across the board with a combination of the best available technologies in order to produce a product with the greatest purity possible. By adopting this approach, pollutant monitoring using chemical fingerprinting methods to ensure that treatment systems are operating properly can used in lieu of drinking water standards.

Because there is a real threat to America's drinking water, this book attempts to show why we need to make a fundamental change in our approach toward protecting this essential resource. In doing so, it presents (1) factual and circumstantial evidence that show the failure of current drinking water standards to adequately protect human health, (2) the extent of pollution in our water resources and drinking water, (3) the currently available state-of-the-art technologies which, if fully employed, would remove pollutants in drinking water to levels much lower than required by today's standards, and (4) the technical means by which we can attain the goal of de- minimus pollution of our drinking water. With this information, the American public will have the means to decide whether or not to demand drinking water of the greatest level of purity possible.

Special thanks go to Jerry Gilbert who provided a constructive review of the text. We owe a sincere debt of gratitude to Paula Massoni for her initial editing of the text and John Kiefer for his technical review and editing. Because we anticipate updating this text on a yearly basis, any reader with relevant drinking water pollution and/or water treatment system performance information is urged to contact the senior author at the Environmental Chemistry and Human Health Institute, 60 E. Third Avenue, San Mateo, California, 94401 or at http://www.psfmaenv@pacbell.net.

Dr. Patrick J. Sullivan
Dr. Franklin J. Agardy
Dr. James J. J. Clark

Chapter 1

Problems

"Our problems would be much simpler if we needed only to consider the balance between food and population. But in the long view the progressive deterioration of our environment may cause more death and misery than the food-production gap."
— Paul R. Ehrlich, *The Population Bomb*, 1971

Drinking water in the United States is polluted with a vast array of man-made chemicals. This condition is the direct result of over 75 years of governmentally sanctioned programs that have permitted low levels of chemical pollutants to be distributed throughout the human habitat. As a consequence of this pollution, drinking water has been and will continue to be polluted by chemicals that are known to be toxic or hazardous at elevated levels but are allowed in our drinking water as long as their concentrations meet existing standards. The safety of these standards is based upon United States Environmental Protection Agency (USEPA) risk assessment models which estimate the probability of human health impacts from the long-term exposure to low concentrations of a specific chemical. However, these models have a key shortcoming. They are all based on data extrapolated from animal studies with no actual calibrated exposure data correlated to their effects on human health. In other words, the real risk to humans is not known.

More importantly, there is an even a larger number of chemical pollutants in drinking water for which no standards have been set. The health effects of the vast majority of these unregulated chemicals on humans are unknown. Because these chemicals are unregulated, the federal government does not impose limits on their concentration in drinking water nor require monitoring for their presence. Furthermore, the effect of this chem-

1

ical soup of regulated and unregulated compounds on human health has never been evaluated nor do any standards exist that regulate the occurrence of chemical mixtures in drinking water.

Historically, the only defense we have against the chemical degradation of our drinking water has been our reliance upon federal drinking water standards. The ability for these standards to actually protect the public health from long-term or even short-term exposure to chemical pollutants in drinking water, however, has yet to be determined. On the other hand, no studies have actually demonstrated that consumption of water meeting all chemical drinking water standards can have harmful results. For some consumers, these omissions and USEPA's claims that our drinking water is safe if it meets their standards are sufficient to allay any fear of harm, but for others it is not. Thus, is it any wonder that Americans are so enamored with bottled water. It certainly isn't because of the economics, especially when bottled water costs as much as 100 to 1,000 times as much as tap water.

As practicing environmental scientists and engineers, we have become increasingly concerned over the failure of present environmental regulations to adequately keep up with the changing nature of the pollutants presently impacting our water resources. It is our belief that our continued reliance upon drinking water standards to protect the public health is a serious mistake because (1) drinking water standards, except for lead, are scientifically unsound and cannot be verified in the real world and (2) the rate at which new chemical and biochemical products are manufactured in the United States greatly exceeds the ability of the USEPA to determine the potential harm of each chemical to humans and establish the necessary standards.

Consequently, our only defense is to abandon our reliance upon federal drinking water standards and adopt a national policy aimed at establishing that all drinking water contain the minimum concentration of chemical pollutants achievable using the best available technology[1] for water treatment. In other words, we must switch from using drinking water standards to protect human health and implement technology-based water treatment requirements to protect human health. Such a policy, however, will demand a fundamental change in how drinking water is treated and distributed in many regions of the United States.

In addition to existing chemical pollution in our drinking water, since the early 1970s there has been a real concern regarding the threat that ter-

---

[1]The term "best available technology" means the best technology that is commercially available.

rorists could poison our drinking water. Sadly, with the September 11th terrorist attacks, the specter of actual chemical or biological attacks on our water supplies appears to be more than just a threat. Given the great difficulty of monitoring for and preventing an unknown toxic chemical or biologic hazard in our water supplies and their post-treatment distribution systems, methods to safeguard against such events need to be implemented.

The chemical threat to our drinking water is real. For the most part, chemical pollution is not an acute threat (i.e., pollution from a chemical spill or terrorist act) but rather a chronic menace in the form of a multitude of chemical pollutants that occur at low concentrations. The health effects of such long-term exposure is virtually impossible to quantify. Such a dilemma fosters inaction and complacency. Yet, the burden of proof should not be placed on the occurrence of "cancer clusters" in various American communities[2] that consume polluted drinking water. In today's chemically dependent society, the use of drinking water standards to protect the public health must be abandoned. To delay will only provide another opportunity to lament a potential future human health tragedy that could have been prevented.

THE BURDEN OF PROOF

Most people, at a gut level, believe that environmental factors, such as the exposure to chemicals, exposure to airborne smoke and particulates, stress, drug and alcohol use, diet, electromagnetic fields and radiation, can contribute to or cause health problems. This belief is strongly supported by an article in the July 2000 *New England Journal of Medicine* [1] that studied the medical history of twins. The article established that it was much more likely that the occurrence of cancer was linked to the exposure of environmental factors than to genetics. These fears are further bolstered by the fact-based films, "A Civil Action" and "Erin Brockovich," which have heightened the public's awareness that chemical pollution[3] of water supplies can have life threatening consequences. In both films, individuals in two communities suffered from illnesses that appeared to be caused by chemical pollution of their drinking water[4] (i.e., trichloroethylene in one case and hexavalent chromium in the other). As a result of these illnesses,

---

[2]For example, Tom's River, New Jersey and Woburn, Massachusetts.

[3]Chemical pollution, for the purpose of this book, is the occurrence of any chemical element or compound above natural background concentrations.

[4]Because pollutants are absorbed through the skin while bathing or inhaled while showering, this water should be as pure as the water we drink.

individuals filed lawsuits against the companies that caused the pollution to recover medical costs and punitive damages for the chemical insult that was unknowingly introduced into their bodies.

In both cases, the sources of chemical pollution and the companies responsible for the pollution were easily determined. The problem facing the attorneys for the plaintiffs was establishing beyond a "reasonable scientific certainty" that there was a direct linkage between the health problems exhibited by their clients and the chemicals they unknowingly consumed. Unfortunately, the attorneys for the plaintiffs in both cases were unable to prove such a linkage. As a result, both cases were settled by mutual agreement between the plaintiffs and defendants with a cash payment to the plaintiffs. Because these cases settled, the courts never affirmed or denied the plaintiffs' allegations that their health problems were caused by the chemicals in their drinking water. From the standpoint of a citizen concerned about the effects of chemical pollution on their health, this is a totally unsatisfying outcome.

These results, however, were virtually preordained. Why? Because, in a court of law, the plaintiffs have the burden of proof. In other words, the plaintiffs must first prove to the court that the company in question actually caused the chemical pollution and secondly that this pollution caused their illness. This second step is not a simple proof. In a toxic tort case, experts for the plaintiffs would have to establish the following sequence of facts:

- What is the pollutant that allegedly caused their illness? This is the simplest fact to establish.
- How did the pollutant reach the location at which the individual was exposed to it? In most cases this can be established with a high degree of certainty.
- What was the concentration of the pollutant at that location where the individual was exposed? This is more difficult, but not impossible, to determine because pollutant concentrations frequently vary with time.
- How long was the individual exposed to the pollutant? While this is not as easily established, reasonable estimates can be made.
- What is the effect of the pollutant on the human body (i.e., its effect at the gene, cell, organ and system level)? This information is known for only a handful of chemicals. Thus, the likelihood of establishing "effect" is nearly impossible, given the tens of thousands of chemicals manufactured in the United States alone.
- For a specific individual, are the symptoms of their specific illness

consistent with exposure to this pollutant? Again, this information is known for only a handful of chemicals for which industrial exposure studies have been completed. Thus, the probability of establishing this relationship in a non-industrial environment is nearly impossible.

Recognizing these weakness, experts for the defense can confuse the issue by emphasizing the potential causative relationship between an illness of a specific individual and any or all of the following environmental factors: exposure to other chemicals either in the home or workplace, smoking, alcohol, diet, stress, drug use, and medical history. It is ironic that other "environmental factors" are used to invalidate the act of actually consuming a chemical pollutant. Defense lawyers will also argue that there is little, if any, scientific evidence linking a specific chemical to a specific health problem[5]. However, one reason for this is that health data (i.e., occupational epidemiologic studies) are inherently inaccurate due to the complexities associated with these studies [2]. Because there usually is no clear relationship between a specific chemical and a human illness, courts often decide to exclude expert testimony on this issue. For example, in Federal Court pursuant to Daubert, expert testimony offered by the plaintiff to demonstrate a link between their illness and exposure to a toxic chemical was found to be inadmissible because it was not scientifically valid [3]. Given this difficult proof, it is no wonder that none of these cases have yet to go to trial.

The failure of plaintiffs to establish the necessary facts in a toxic tort case is the direct result of not having chemical-specific information on the relationship between exposure to known concentrations of a chemical and human health effects[6]. The only chemical for which these relationships are fairly well known is lead and this is only because the health effects of lead on humans have been extensively studied for nearly a century. In today's litigious environment, the ability to develop human specific chemical toxicity data on the tens of thousands of man-made chemicals used in the United States would appear to be impossible unless the federal government allows human chemical exposure studies. Since this very highly unlikely,

---

[5]In another Erin Brockovich/PG&E chromium lawsuit in Kettleman, California reported in the San Francisco Chronicle on January 12, 2002, PG&E spokesman Jon Tremayne stated that the defense "will be to question the links between the chromium contamination and the illnesses of plant workers and local residents."

[6]This can be generally referred to as a dose-response relationship. This relationahip would quantify the specific chemical concentrations required to induce a specific health response (e.g., hormone imbalance, cancer, death, etc.).

researchers will never be able to evaluate the true impact of specific chemical pollutants on human health, let alone the effect of mixtures of these chemicals that occur in drinking water.

Because chemical toxicity is only based on animal studies, the concentration at which a specific chemical causes a toxic response or cancer in a human is not known. This failing calls into question the ability of the government to establish the concentration, or standard, at which a chemical in drinking water will not impact human health. After all, what is the purpose of establishing a "safe" drinking water standard if it is not based on actual human exposure data.

If it is not possible to demonstrate a causative link between a chemical and harm to the human body in a court of law, then there is no assurance that drinking water is safe based on arbitrary standards. In spite of this critical deficiency, existing federal laws imply that there is "no harm" if a chemical pollutant has a concentration that is less than an existing water quality standard. This dichotomy has perpetuated a federal policy of legalized pollution[7] that is advertized as being protective of the public health. Thus, according to federal and state governmental programs, its acceptable to (1) drink small quantities of chemicals known to have harmful effects to animals at elevated concentrations, (2) drink any concentration of a known or potentially toxic chemical that does not have a federal or state standard, which is approximately 99.5 percent of all the chemicals currently used in the United States, and (3) drink water containing a mixture of chemical pollutants just as long as each chemical is either unregulated or below its published standard.

A LICENSE TO POLLUTE

Since the 1700s, rivers have been the natural extension of municipal and industrial wastewater discharge systems (i.e., wastes were discharged directly into lakes and rivers). With increasing population, pollution increased to the point that by the late 1800s vocal public concern forced the development of state regulations to control pollution. By 1904, laws in the United States prohibited water pollution [4]. As a consequence, there was no immediate need for water pollution standards. Sadly, this condition did not last very long.

The evolution from "no pollution" at the state level, to "acceptable levels of pollution" defined at the federal level, occurred in the innocent form

---

[7]Federal and state environmental laws allow industry and/or individuals to release chemicals to water as long as the pollution is below "acceptable" levels.

of the national water quality standards[8] that were published in 1925. The objective of the U.S. Public Health Service (USPHS) in publishing drinking water standards [5] was to "safeguard the health of the public" against the most seriously recognized dangers such as "contamination by disease such as typhoid fever and other illnesses of similar origin and transmission." These standards did, however, warn that "to state that a water supply is 'safe' does not necessarily signify that absolutely no risk is ever incurred in drinking it." This statement, which was made more than three quarters of a century ago, is probably one of the most accurate predictions of today's pollution problems.

The USPHS water quality standards further stated that an acceptable water supply should be suitable for drinking, be clear, colorless, odorless, pleasant to the taste and free from toxic compounds. These standards also established limits for the following toxic metals in drinking water: lead at 0.01 parts per million (ppm)[9], copper at 0.2 ppm and zinc at 5.0 ppm. If a water resource exceeded any of these standards it was to be "rejected" and an alternate source used in its place. Unlike 1925, however, today it is more difficult to find an alternate water source that is not polluted.

By setting specific chemical standards with defined concentrations that are assumed to be safe, the federal government established a policy of "permissible pollution." In other words, the federal government essentially issued a license to allow drinking water to be polluted up to an acceptable level[10]. As a natural consequence of setting drinking water standards, specific definitions for the terms "contamination" and "pollution" were developed and these are presented in Exhibit 1.1.

Chemical pollution is "licensed" or permitted for industrial and commercial businesses. These regulated sources of chemical releases are referred to as point sources of pollution. Under current federal and state law, regulated pollutants can be discharged into local surface waters up to a specified limit, while other unregulated chemicals can be discharged with no specific chemical limitations. The other type of chemical pollution that occurs is from non-point sources. Non-point sources of chemical pollution are generally distributed over a large geographic area and are much more dif-

---

[8]Given that surface and groundwater resources cross state boundaries, a national water quality policy was necessary.

[9]In water, ppm is expressed as milligrams of a chemical per liter of water or mg/l. For example, 1 ppm of sugar dissolved in water is the equivalent to mixing a five pound bag of sugar into an Olympic size pool.

[10]This policy of licensing pollution continues today and is referred to in this text as "standard based pollution."

Exhibit 1.1 Definitions

Because water quality standards set acceptable levels of pollution based on their potential for health effects, some states have defined pollution based on the knowledge of health hazards. For example, the State of California has used the following definitions since the 1960s:

- Contamination is defined as any impairment of the quality of the water of the State by sewage or industrial waste to a degree which creates an actual hazard to public health through poisoning or through spread of disease.
- Pollution is defined as an impairment of the quality of the waters of the State by sewage or industrial waste to a degree that does not create an actual hazard to public health but does adversely and unreasonably affect such water for domestic, industrial, agricultural, navigational, recreational, or other beneficial use.
- Nuisance is defined as damage to any community by odors or unsightliness resulting from unreasonable practices in the disposal of sewage or industrial waste.

These definitions clearly define the difference in the terms contamination and pollution as they relate to public health. But since there is no actual demonstrated basis for human health standards, the term "contamination" is meaningless. Yet all of the Federal clean water programs use of the term "contamination" to define the threat to public health. That is why we describe the impacts on our water resources as "pollution."

Source: McKee, J. E. and H. W. Wolf, "Water Quality Criteria," California State Water Quality Control Board, Publication No. 3-A (1963).

ficult to control. As a result, many non-point sources cannot be regulated like point sources of pollution. Examples of both point and non-point sources of pollution are given in Exhibit 1.2 (*see* pp. 10–11). As a result of these sources of pollution, a vast array of chemicals are distributed throughout surface water and ground water.

The types of chemicals released to the environment have evolved since the first standards were implemented in 1925. These chemicals are not only more complex but there are many more of them. It has been estimated that in excess of 72,000 chemicals are produced in the United States [6], yet the number of these chemicals that are currently regulated is extremely small. In addition to the chemicals discharged into the environment, the process of purifying water using chlorine and bromine also creates disinfection by-products (DBPs) in treated drinking water and wastewater. Some of the DBPs of concern that occur in water supplies and bottled water include trihalomethanes, haloacetic acids, chlorite, bromate and N-nitrosodimethylamine. A very small number of these chemicals are regulated under the current drinking water standards.

The current federal Primary National Drinking Water Standards for chemicals are listed in Appendix A. The primary standards are those that are legally enforceable and apply to public water systems[11]. The January 2002 list includes 63 organic compounds (i.e., chemicals primarily composed of carbon) and 19 inorganic compounds. For each chemical on the list, the standard is expressed as a maximum contaminant level (MCL) that should not be exceeded[12]. If the level is exceeded, consumers are to be warned by the public water system that their water has been polluted by a specific "regulated" chemical,[13] but there is no regulatory requirement to remove that chemical. In some cases, water utilities will implement, at a substantial cost to the consumer, advanced treatment methods to remove pollutants. For example, because of the widespread pollution of drinking water by pesticides in Delaware, Iowa, Indiana, Kansas, Maryland, Michigan, Minnesota, Missouri, Nebraska and Ohio, some water utilities in this "corn belt" region have had to upgrade their water treatment facilities [8]. For consumers with polluted drinking water, they must either find another source of drinking water or treat the water in their own home until the problem is corrected by the water utility[14].

Of the 168,690 public water systems [7] in the United States, only 7.6 percent of these have actually monitored their water for the chemicals in the primary Drinking Water Standards list and reported the occurrence of these chemicals to the USEPA. These data are compiled in the USEPA's National Drinking Water Contaminant Occurrence Database and can be accessed at http://www.epa.gov. Of the public water systems that did report

---

[11]A public water system is a system for the provision to the public of water for human consumption through pipes or other constructed conveyances, if such system has at least 15 service connections or regularly serves at least 25 individuals at least 60 days out of the year. Thus, there are no regulations covering private water systems which account for almost 20,000,000 Americans [7].

[12]Asbestos is the 20th inorganic compound but the standard is in fibers per liter.

[13]More importantly, when an unregulated chemical pollutant (i.e., a chemical without a primary drinking water standard) is identified in drinking water no warning is required.

[14]In specific cases where a industry or commercial business was responsible for groundwater pollution of drinking water sources, the polluting party will usually be required to pay for the damage to the resource (e.g., supply bottled water to consumers).

Exhibit 1.2 Point and Non-point Sources of Chemical Pollution

When attempting to evaluate the magnitude and extent of pollution on water resources, it is critical to distinguish between point and non-point sources of pollution. Point source pollution is typified by the discharge of chemicals in either gas, liquid and/or solid form from a facility that has an identified point of release. For example:

- Chemicals are released into the atmosphere in the form of aerosols, gases and/or particulates from stacks, flares, conveyors and towers.
- Chemicals are discharged into surface water (rivers, lakes, bays and oceans) from industrial wastewater and sewage outfalls.
- Chemicals are released onto the land in the form of wastewater irrigation or injected into deep groundwater aquifers.

Pollutants released into the atmosphere have the potential to travel extremely long distances and can subsequently be deposited in a surface water resource. Wastewater, and any solids contained therein, is an obvious source of pollution to surface water. However, it should not be forgotten that surface water, in all but the most rare circumstances, will recharge groundwater. In a similar fashion, chemicals placed into or on land may leach into both surface water and groundwater.

Non-point source pollution is typified by processes that result in the release of chemicals across a fairly broad or diffuse geographic area. For example:

- The extraction of coal and metal sulfides (i.e., gold, silver, copper, lead, zinc, etc.) from the earth can disturb thousands of acres of land. Because of this disturbance, trace metals and acid can be released from areas of exposed coal and minerals (e.g., high walls, overburden, shafts, adits, waste rock piles, tailings and mineral processing waste piles).
- The widespread application of pesticides to agricultural lands can pollute irrigation water, surface water runoff from rainfall and ultimately creeks, rivers, ponds, lakes, oceans and groundwater.

(Exhibit 1.2, conrinued)

- Rural and agricultural areas used to feed livestock are a source of nitrogen, phosphors, hormones and antibiotics to both surface water and groundwater.
- Urban environments are subject to the accumulation of chemicals on buildings, streets, and impervious surfaces, as well as, in storm channels and sewers and areas of waste storage or disposal. Common pollutants include pesticides, nitrogen, phosphorus, metals, solvents and petroleum products. During rainfall events, these chemicals will be washed off and carried in the runoff to local receiving waters.
- Nitrogen and phosphorus can be found in the runoff from agricultural lands and residential lawns can also contribute these same pollutants to surface waters.
- Certain rocks and ore bodies are sources of nitrogen, perchlorate, fluoride and radioactive elements that can pollute groundwater.
- Pollutants that have been deposited in creek, river, lake and ocean sediments in the past, may continuously leach into water or be released when the sediment is disturbed (i.e., by erosion or dredging).

Another category of non-point pollution sources are chemical releases from industrial facilities, commercial businesses and waste disposal sites that either manufactured, handled, stored or disposed of hazardous chemicals and wastes. Although pollution from such facilities may have a general point source location, the chemical pollution to either surface water and/or groundwater can be spread over a wide geographic area.

The distinction between point and non-point sources of pollution is important since our ability to prevent a point source of pollution from entering a water resource is much easier than preventing pollution from non-point sources because of the diffuse nature of the pollution. Although the USEPA has begun to address runoff from urban and agricultural landuses and mining, there are no quick economical or easy solutions on the horizon. Regardless of the level of point source control, widespread pollution of water resources from non-point sources will continue to exist for the foreseeable future.

to the USEPA, approximately 90 percent reported the detection of at least one chemical from the list as being present in their drinking water. In many cases, the standards for these chemicals were exceeded. Unfortunately, "standard-based pollution" has been adopted throughout federal regulations [9].

Every significant pollution control regulation in the United States (e.g., Clean Water Act, Safe Drinking Water Act, Clean Air Act, Resource Conservation and Recovery Act and the Comprehensive Environmental Response Compensation and Liability Act) allows for chemical pollution to set limits that are established to "protect human health and the environment." Because there is practically no scientific data showing a direct relationship between the adsorption and ingestion of specific chemical pollutants or mixtures of chemicals and observed human health problems, the pollution control regulations that have been in effect for the last 30 years cannot be shown, in a demonstrable sense, to protect human health. Without chemical- specific and health-specific information, the entire basis of pollution control regulations in the United States could be viewed as a travesty or, at best, a failed attempt at safety.

Furthermore, even when human health risks have been demonstrated for specific chemicals, it usually takes the federal government years to recognize, let alone address the hazard. For some of the most common water pollutants, the gap between the time a compound has been released into the environment and its recognition as a health hazard by the government can be decades. Such delays are not only unacceptable, but also result in extensive and often irreversible damage to the nation's water resources. Only time will tell if similarly irreversible damage has been caused to humans.

TIMING IS EVERYTHING

The delay in recognizing impact of chemicals on human health is another layer of risk that is intrinsic to a "standard-based" pollution control policy. For example, such time delays were characteristic of hexavalent chromium[15], which was the chemical of concern in "Erin Brockovich," and trichloroethylene, which was the chemical of concern in "A Civil Action". Chromium has been known as a hazardous chemical since the 1850s,

---

[15] A form of chromium that is characterized by its electrical charge or valance. Hexavalent chromium has a charge of +6 and is more toxic than trivalent chromium (+3). Hexavalent chromium in the environmental is almost always man-made while trivalent chromium occurs naturally and is a common constituent in vitamins. The analysis of chromium is reported as either Total Chromium (i.e., hexavalent + trivalent) or as Hexavalent Chromium.

but not until the1930s were there wide-spread reports of the toxicity of chromium to both aquatic life and humans [10, 11, 12, 13 and 14]. By the1940s, the pollution of both surface water and groundwater by chromium was a frequent occurrence. Due to this hazard, the USPHS in 1946 set a water quality standard of 0.05 ppm for hexavalent chromium. Even though chromium was a known toxic compound, when the first drinking water standards were set in 1925, it took another 21 years to add chromium to the list. Moreover, the current standard of 0.1 ppm is for total chromium and does not even include a standard for hexavalent chromium, which is suspected to be far more toxic. Unlike the federal government, the State of California has proposed a public health goal for total chromium in drinking water at 2.5 ppb and initiated cancer studies on hexavalent chromium. The concern for hexavalent chromium is in no small part the result of a California drinking water survey in September 2001 that revealed 375 drinking water sources contained hexavalent chromium at levels greater than 5.0 ppb [15]. Changes of this sort at both the federal and state level typify the capricious nature of setting water quality standards and cast doubt on the scientific validity of these modifications.

Trichloroethylene (TCE) is an organic compound that contains chlorine [16]. TCE has been used as an organic solvent (i.e., it dissolves other organic compounds like oil) since the 1920s. Because of its widespread use in industry, most aspects of TCE toxicity were established in the 1930s. Beginning in the late 1940s, TCE was identified as an environmental pollutant. By 1975, TCE was identified as a possible carcinogen. Even with this knowledge, it took the until 1987, or almost 40 years, for USEPA to establish a final maximum contaminant level in water of 0.005 ppm [17].

These two chemical examples, chromium and TCE, represent two of the most significant chemical pollutants of the past 75 years. Unfortunately, we still face the similar time delays as scientific data trickle in on only a very small number of the thousands of compounds that are released daily into our environment. In almost all cases, these chemicals do not have established federal drinking water standards.

Examples that typify the ongoing delays in USEPA's recognition of hazardous chemicals in the environment center on (1) methyl tertiary-butyl ether (MTBE)[16], (2) perfluoro-octanyl sulfonate (PFOS), (3) perchlorate, and (4) dioxin. In the late 1970s, MTBE was added to gasoline as an octane-enhancing compound. Prior to its introduction into gasoline,

---

[16]Today, a common groundwater pollutant reported nationwide that usually requires costly remedial action.

the Toxic Substance Control Act Interagency Testing Committee, an independent advisory committee to the USEPA, recommended that MTBE's toxicity be evaluated. Toxicity data (genetic, reproductive and carcinogenic) developed in the early- to mid-1990s resulted in a 1997 recommendation that MTBE exposure be limited to 0.01 to 0.02 ppm [18]. Thus, it took approximately 20 years after MTBE was added to gasoline for it to be identified as a chemical hazard by the USEPA. In spite of the toxic nature of this chemical, MTBE has yet to be added to the current primary drinking water standards. This is a serious omission since a January 2000 study by the American Water Work Association [19] reported that, "MTBE contamination of drinking water supplies is fairly widespread and may occur in almost any area where gasoline is used." What is more disturbing, however, is that this same article reports that existing water treatment methods do not remove MTBE. Since MTBE is currently unregulated, this means that MTBE polluted drinking water can be distributed directly to consumers without so much as a warning.

PFOS is the primary active ingredient of 3M Corporation's Scotchgard and is also incorporated into microwave popcorn bags and fast-food wrappers. As of 2002, 3M Corporation will cease production[17] of PFOS because it has been discovered in the blood of humans and animals in pristine geographic areas of the world "where no apparent sources exist." This compound is now considered, approximately 40 years after its production began, a significant chemical hazard to humans and the environment based on its known toxicity and its extremely long persistence [20]. This chemical, like MTBE, is also unregulated.

Perchlorate has been widely used since the late 1940s in rocket fuel and lubricating oils, for the tanning and finishing of leather, as a fixer for fabrics and dyes, in electroplating, aluminum refining, rubber manufacture, and paint production. Perchlorate was not a chemical of environmental concern until 1997 when a new analytical method enabled scientists to detect it in water at concentrations as low as 4 ppb[18]. Using this new method, California water quality agencies found perchlorate in 140 public water supply wells. Because there was no federal drinking water standard for perchlorate, California ruled in 1997 that drinking water should not

---

[17]PFOS production is less than 10,000 pounds per year.

[18]In water, ppb is expressed as micrograms of a chemical per liter of water or ug/l. For example, 1 ppb of sugar dissolved in water is the equivalent to mixing a packet of sugar (approximately 3 grams) into an Olympic size pool.

[19]No one can agree on the appropriate standard. Arizona set a provisional health based level at 31 ppb while Texas set an interim level at 22 ppb.

contain more than 18 ppb of perchlorate [21][19]. In January 2002, California's Department of Health Services reduced the perchlorate Action Level to 4 ppb. Thus, some 50 years after its introduction, the State of California, but not the USEPA, finally recognized that perchlorate is a hazard in drinking water. This is because perchlorate is one of the many "endocrine-disrupting" chemicals that are not currently regulated by the USEPA, even though these compounds are believed to increase the risk of testicular, prostate and breast cancer in humans [22].

Dioxin, a chlorinated organic chemical identified as 2, 3, 7, 8-tetra-chlorodibenzo-p- dioxin (TCDD), received its initial notoriety in 1972 as the result of the first true environmental disaster in United States history. A chemical division of Syntex Corporation in Verona, Missouri gave a waste oil containing dioxin from the production of hexachlorophene (a common disinfectant) to a Mr. Russell Bliss. Mr. Bliss in turn sprayed the waste oil in horse arenas and on dirt roads throughout southeastern Missouri to control dust. The number of horses that died following the spraying were so numerous that they had to be mounded in piles and burned at several separate arenas. Mr. Bliss also sprayed the same waste oil on all the roads in Times Beach, Missouri. Once this was discovered, the entire town had to be evacuated and was eventually purchased by the federal government. The total cost of the Missouri dioxin cleanup exceeded 1 billion dollars. At that time, the Center for Disease Control proclaimed dioxin to be the most toxic compound known to man.

Since this incident and at about ten-year intervals, reports on the toxicity of dioxin have ranged from its being either extremely toxic or not really being a problem. As of the September 2000 Draft Dioxin Reassessment report, dioxin is now listed as a "known" human carcinogen. Regardless of its level of toxicity, the fact remains that dioxin is still allowed in drinking water[20] at an "acceptable level." Once again, it has taken the USEPA nearly 30 years to decide if a chemical is toxic to humans. Such delays remain inherent to the USEPA programs that attempt to evaluate specific chemical hazards.

Delays by the USEPA, however, should not be surprising since the National Research Council estimates that there are approximately 72,000 organic chemicals[21] in commerce within the United States, with nearly 2000 new chemicals being added each year [6]. Given this number of chemicals, it would seem impossible that the USEPA could identify every

---

[20]As of March 2001, only 71 public water supplies (out of 168,690) have been tested for dioxin.
[21]Either produced or imported by facilities in greater than 10,000 lbs/year quantities.

chemical that poses a hazard to the public health. In fact, it is impossible. For example, under the Toxic Substances Control Act that was established in 1979, the USEPA has conducted an assessment program to determine which "new" chemicals present an unreasonable risk to human health or the environment. Since 1979, the USEPA has only reviewed approximately 32,000 new chemical substances. With so many chemicals left to evaluate, the USEPA chose to exclude all chemicals that are produced in amounts less than 10,000 pounds per year and all polymers[22] from further consideration. "The remaining 15,000 chemical subset has been identified as being the broad focus of the USEPA's existing chemical testing and assessment program with the primary focus being on the 3000 high production volume chemicals that are produced/imported at levels above 1 million pounds per year [6]." Thus, the USEPA has limited the number of chemicals it will consider as having an "unreasonable risk to human health and the environment" without even considering toxicity. The error of such a selection process has already been demonstrated by the threat posed by perfluoro-octanyl sulfonate (as discussed above). Because USEPA has adopted this approach to assessing which chemicals are a threat to human health, many chemicals that are truly hazardous will not be evaluated. As a result, the USEPA will never be able to collect the necessary toxicity data to determine which chemicals need to be regulated in our drinking water resources.

Once toxicity data are collected from animal studies, the USEPA must still identify those chemicals that should be regulated in drinking water. As of 1998, USEPA had completed their evaluation of approximately 400 compounds for potential regulation in drinking water. These "top 400" chemicals were coalesced into one master list from existing lists of chemicals that were already incorporated into various regulations. Unfortunately, the "top 400" were selected independently from the "top 3000" chemicals identified by USEPA's toxicity evaluation program. In other words, the USEPA failed to consider the inclusion of any of the "top 3000" chemicals into its initial list of chemicals to be considered for potential regulation in drinking water. This omission was pointed out by the National Research Council [6] as a major flaw in the USEPA program to select chemicals for addition to the drinking water standards since not all chemicals of environmental concern were included for consideration.

This issue aside, the USEPA further reduced the "top 400" to a final list

---

[22]These are high molecular weight chemicals which are assumed to typically exhibit low toxicity and water solubility.

of 50 chemicals that is defined as the "Drinking Water Contaminant Candidate List"(see USEPA's Contaminant Candidate List in Appendix B). Of the final 50 chemicals on this list, a minimum of 5 compounds must be selected for regulation under the primary drinking water standards within five years [2, 6]. This is a disturbing result. From a scientific and practical point of view, the USEPA never had either the time or resources to comprehensively evaluate and regulate the entire list of 72,000 chemicals. However, to select a minimum of five chemicals in five years for regulation is just as absurd. It is extremely difficult to accept the premise that of all the chemicals used in this country there are only five additional chemicals that are a hazard to human health. Yet, the USEPA expects us to believe that by regulating an additional five chemicals our drinking water will be safe. This is particularly egregious since the methods employed by the USEPA for choosing these chemicals was flawed from the very beginning.

Consequently, many hazardous chemicals that should be regulated on the basis of their potential toxicity to humans will go unregulated and when these unregulated chemicals occur in drinking water there will be no requirement to warn the consumer[23]. Furthermore, based on the USEPA's performance to-date and their pace of evaluating and regulating chemicals in drinking water, we might all be dead by the time the USEPA ponders the fate of its first list of 50 chemicals. Such delays are inherent in current federal policy. For example, the Toxic Substances Control Act requires the USEPA demonstrate that a chemical is dangerous before it can take any action against that chemical (i.e., regulate that chemical). It is because of this type of policy that the presence of unregulated chemicals in our drinking water represents one of the greatest threats to America's public health. A perfect example of this problem is typified by chemicals identified as "endocrine-disrupting chemicals."

ENDOCRINE-DISRUPTING CHEMICALS

The endocrine system regulates bodily functions through hormones released from the brain, thyroid, ovaries, testes and other endocrine glands. Many chemicals are now being identified as substances that can interfere with normal hormone function. These endocrine- disrupting chemicals can effect reproduction and can increase a person's susceptibility to cancer and other diseases evan at low ppb levels of exposure. Given this hazard, the

---

[23]Community water systems are only required to notify consumers if regulated pollutants exceed their established criteria.

USEPA has proposed a Endocrine Disruptor Screening Program that will gather the information necessary to identify those endocrine-disrupting chemicals that should be regulated. In the USEPA's report to Congress in August of 2000 [23], it estimated that approximately 87,000[24] chemicals currently in "commerce" would have to be evaluated to define their potential risks. As a first step to the USEPA process, the USEPA has proposed to focus on finding methods and procedures to detect and characterize the effects of pesticides, commercial chemicals and environmental pollutants on endocrine activity. Their proposed schedule for selecting valid testing methods will take about four years. Obviously, actual testing using the approved methods will take much longer. Given this type of schedule, it will be decades before any specific endocrine-disrupting chemicals are identified by the USEPA as being a hazard to human health in general. Determining whether these chemicals should be allowed in drinking water will take even longer.

Thus, the vast majority of endocrine-disrupting chemicals will remain unregulated for a very long time. Although many of these compounds have not yet been identified, a general list of suspected endocrine-disrupting chemicals are presented in Appendix C. Of these suspected endocrine disrupting-chemicals, 19 are regulated under the Primary Drinking Water Standards as being hazardous for other reasons (i.e., independent of any determination of their endocrine-disrupting characteristics). Because 93 chemicals are suspected of disrupting endocrine functions, this leaves another 74 chemicals unregulated. Eight of these compounds are on USEPA's Contaminant Candidate List and may be regulated in the next five years. How can the remaining 74 chemicals not be a potential health risk? We may not know the answer to this question until we are well into the 21st Century. Sadly, the American public is left to assume that if any of these unregulated chemicals occur in drinking water, then they are not a threat to human health. This is an assumption that is hard to swallow.

In the meantime, an article by the American Water Works Association [24], recommends that "the utility [water provider to a community] should take measurements to establish whether compounds at issue [endocrine-disrupting chemicals] are present in the untreated water and how effective treatment is at reducing compound concentrations." (In the case of

---

[24]This number is significantly larger than the 72,000 chemical reported by the National Research Council [6] in 1999.

endocrine disruptors, the concentrations are so low that conventional analytical techniques also may not detect them.) Such actions will not solve the problem. However, if water utilities begin this process, at least the problem will become better defined.

As bad as this problem is, it is even more alarming that, for the most part, the current methods for evaluating toxicity are limited to individual chemical studies. Virtually no scientific data on the toxicity hazard posed by mixtures of both regulated and unregulated compounds currently exist. Furthermore, a whole class of chemical compounds is not even being tested by the USEPA. These compounds include the hormones, antibiotics and synthetic organic compounds that are manufactured by the pharmaceutical industry.

PHARMACEUTICAL POLLUTANTS

In the early 1980s, seminars conducted by the American Association for the Advancement of Science, in cooperation with the USEPA, reported that mixtures of common pharmaceuticals (i.e., female hormones, tranquilizers and diuretics) were widely and routinely detected in water supplies. The hazard posed by these compounds in low but chronic levels was not then known nor is it known now (some of these compounds can also be endocrine disruptors). Additionally, the number of medications entering into and persisting in our water resources has increased. Currently, public water supplies do not monitor or report to the USEPA on the occurrence of pharmaceuticals in drinking water. This raises a question as to the degree to which pharmaceuticals pollute our water resources as well as their ultimate effect on the consumer.

Medications that pass through the human body and unused drugs that are dumped into sinks and toilets flow through sewers to publicly operated treatment works (POTW). Once in the POTW[25], these compounds have been shown to survive the treatment process and ultimately end up being discharged to both surface and groundwater [25]. A 1998 report on drugs found in the effluent from POTWs identified more than 20 pharmaceuticals [26]. Two of these drugs were clofibrate (a cholesterol-lowering drug) and cyclophosphamide (a drug used in chemotherapy). A complete list of these compounds is given in Exhibit 1.3. Without monitoring for these chemicals, how do we determine their presence or absence and, if present,

---

[25]The purpose of a POTW is to treat sewage to destroy bacterial hazards and reduce chemical pollutants prior to the waste being discharged back into the environment.

Exhibit 1.3 Pharmaceuticals in Water Resources

According to the article by Tara Hunt in the July 2000 issue of *Water Environment and Technology*, the following pharmaceuticals have been found in the environment:

Type of Drug	Human Drug	Veterinary Drug
Analgesic	Aspirin	
	Dextropropoxyphene	
	Dichlorfenac	
	Ibuprofen	
	Indometacin	
Antibiotic	Erythromycin	Oxytetracycline
	Phenicilloyl groups	
	Sulphamethoxazote	
	Tetracycline	
Antiparacitic		Ivermactin
Anxiolytic	Diazepam	
Cancer treatment	Bleomycin	
	Cyclophosphamide	
	Ifosfamide	
	Methotrexate	
Hormone	Estrogen	Estrogen
	Estradiol	Testroesterone
	Estrone	
	Ethinylestradiol	
	Norethisterone	
	Oral Contraceptives	
	Testosterone	
Cholesterol-lowering	Clofibrate	
	Clofibric acid	
Narcotic	Morphinan-structure	
Psychomotor	Caffeine	
	Theophylline	

(Exhibit 1.3, continued)
Most researchers and government officials interviewed have said that more research is required to determine the extent of any pharmaceutical pollution, but that finding the necessary financial support to do so would be difficult. In addition, the ability to control such pollution requires that the source be identified. Currently, there are three potential sources: the chemical manufacture, medical facilities that distribute pharmaceuticals, and the patients that receive and use them. Ultimately, some portion of waste generated by all of these sources may end up at a public sewage treatment plant. As a result, municipal wastewater treatment plants may become responsible for the removal of pharmaceuticals from wastewater discharges. If this occurs, the cost of treating wastewater to remove pharmaceuticals could increase treatment costs by at least 10%. Given this cost, wastewater treatment engineers argued that it would be more cost effective to remove pharmaceuticals from drinking water instead. Regardless of where the treatment is installed, "until now we've only seen the top of a large iceberg...many compounds, especially some metabolites, cannot be detected or extracted in water using our current methods."

in what concentration? POTWs do not monitor for these compounds in their discharges.

In 1999, the United States Geological Survey [USGS] did begin a program to monitor pharmaceuticals, including prescription and non-prescription drugs as well as sex and steroidal hormones, in surface and groundwater. This monitoring under the Toxic Substances Hydrology Program was expanded to 95 chemicals in 2000 [27]. These compounds, including some industrial, insecticides and home care product chemicals, are listed in Appendix D. The USGS study sampled 139 rivers and streams in 30 states for the list of 95 chemicals of which only 14 have established water quality standards. The most frequently detected compounds included coprostanol (fecal steroid), cholesterol (plant and animal steroid), N-N-diethyltoluamide (insect repellant), caffeine (stimulant), triclosan (antimicrobial disinfectant), tri(2-chloroethyl)phosphate (fire retardant) and 4-nonylphenol (detergent metabolite). Concentrations of the detected compounds were generally less than 1 ppb. In half of the streams sampled, seven or more compounds were detected and in one stream 38 chemicals were present. The researchers also stated that limited information is available on the potential health effects to human and aquatic ecosystems from low-level, long-term exposure to these chemicals or combinations of these chemicals. One can only wonder as to the number of chemicals that will be monitored in the following years and what chemicals will be left out of the process. This type of research truly demonstrates that a problem exists, but offers no solution.

According to the *Journal of the American Water Works Association* [28], concern is growing over pharmaceuticals finding their way from medicine cabinets and sewage into the nation's water supplies and no one is clear on what the cumulative effect of this onslaught will turn out to be. As a result, it might be wise to avoid making facile generalizations about the safety of our drinking water.

In the meantime, the ability to prevent prescription and non-prescription drugs as well as sex and steroidal hormone pollution from POTWs will require both better controls on discharges and additional wastewater treatment prior to discharge of effluents to the environment. Such controls, however, assume that the government will require the removal of these unregulated chemicals from wastewater. Since wastewater treatment costs would increase an estimated 10 percent [27], additional funds will be needed to limit this type of pollution. As a result, either the consumer will have to pay for the increased cost of treatment or the state and federal government must provide the funding. Until the public demands a change, inadequately treated wastewater will continue to be discharged into the country's water resources, polluting our drinking water with a vast array of chemicals.

These polluted waters often serve as the water resource for a downstream public water supply. Although most public water systems provide water treatment, the standard methods employed do not necessarily remove all chemical pollutants from the water. When this deficiency is combined with the fact that public water systems do not even monitor for these compounds, consumers are exposed to an unknown pollution risk. This risk cycle needs to be broken.

A governmental policy that relies on water quality standards has left the public vulnerable to the hazards of both regulated and unregulated pollution. For example, a 1995 study by the United States Geological Survey, which is summarized in Exhibit 1.4, illustrates the extent of organic chemical pollution in the Mississippi River. Given the wastewater cycle of permitted sources of pollution (such as from POTWs), it is no wonder that cities such as New Orleans, which is located at the end of one of the longest wastewater/water cycles, the Mississippi River, have elevated cancer rates [29]. This problem was most succinctly addressed in an article of the

Exhibit 1.4 Mississippi River Report

A 1995 report by the United States Geological Survey (Circular 1133) presents data on "Organic Contamination of the Mississippi River from Municipal and Industrial Wastewater." According to this report, "The Mississippi River receives a variety of organic wastes, some of which are detrimental to human health and aquatic organisms. Urban areas, farms, factories, and individual households all contribute to contamination, by organic compounds, of the Mississippi River. This contamination is important because about 70 cities rely on the Mississippi River as a source of drinking water." This conclusion was based on a water sampling program that was conducted between Minneapolis-St. Paul, Minnesota, and New Orleans, Louisiana from 1987 and 1992.

These water analyses show that the Mississippi River is contaminated with the following groups of compounds:

- Methylene-blue substances from synthetic and natural anionic surfactants (e.g., detergents).
- Linear alkylbenzenesulfonates, a complex mixture of anionic surfactant compounds used in soap and detergent products.
- Nonionic surfactants such as nonylphenol and polyethylene glycol.
- Adsorbable halogen-containing organic compounds including solvents and pesticides.
- Polynuclear aromatic hydrocarbons from the combustion of fuels.
- Caffeine from beverages, food products and medications.
- Ethylenediaminetetraacetic acid (EDTA), a widely used synthetic chemical for complexing metals.
- Volatile organic compounds including chlorinated solvents and aromatic hydrocarbons.
- Semivolatile organic compounds including priority pollutants such as trimethyltriazinetrione (a by-product of methylisocyanate) and trihaloalkylphosphates (a flame-retardant).

This list is considerably different from the USEPA Primary Drinking Water Quality Standards presented in Appendix A and the chemicals listed in Exhibit 1.6. These differences further highlight the impossible task of defining which chemicals should be selected for monitoring, let alone which chemicals should be regulated.

Journal of the American Water Works [30] that stated "...the boundaries between water and wastewater are already beginning to fade. For example, on some major rivers in the United States, water is used and reused up to 20 times as it travels to the sea—the discharge water from one wastewater treatment plant comprising the raw water intake for a primary drinking water plant a few miles downstream." Advocates [31] that support the reuse of wastewater[26] contend that existing projects have "operated safely and reliably for nearly 40 years." This statement, however, only refers to bio-logic pollution and not to chemical pollution since these programs do not monitor for unregulated pollutants.

In addition to pollution arising from human use of pharmaceutical products, agricultural use of antibiotics is also a significant source of pollution. According to an August 2000 report from the Environmental Defense Fund to the USEPA, approximately 40 percent of all antibiotics used in the United States are used in animal production. Furthermore, as much as 80 percent of the antibiotics administered orally pass through an animal unchanged. As a consequence, antibiotics pollute both surface and ground water resources.

The pollution of water resources from both human and animal sources is well documented. Yet, no pharmaceuticals are currently being evaluated by the USEPA as a threat to drinking water. This condition continues to highlight the failure of federal water quality standards to effectively define chemical risks and protect human health within a reasonable time frame. In addition to the problems manifested by the USEPA efforts to define drinking water standards, the specific standards that are ultimately enacted at the state level are dependent upon individual state regulatory agency decisions. While states must at a minimum meet federal drinking water standards, state standards may be more strict. Thus, state regulatory decisions are also a concern as is illustrated in Exhibit 1.5 (*see* pp. 26–27).

LIVING WITH RISK

Because we accept federal standards, consumers of either public water supplies or bottled water are forced to live with the following unknown risks:

1. What are the real health risks associated with consuming regulated chemical compounds at or below their established standards?

---

[26]Treated wastewater is usually injected into a groundwater aquifer where natural adsorption of pollutants is assumed to occur. Also it is assumed that the majority of clean water in the aquifer will dilute the pollution. These assumptions, however, are not always correct.

2. What unregulated pollutants occur in drinking water and what are their concentrations and health risks?

3. What are the actual concentrations of regulated compounds in drinking water[27]?

4. What effect do low levels of a multitude of pollutants have on the human body?

Given the current history of water pollution control in the United States, we will continue to suffer "legalized chemical pollution" until science demonstrates the actual nature of the harm or until there is a change in policy. Until that time, should we be expected to accept this risk[28] based on statistical assurances of safety? A blind acceptance of this risk is difficult to swallow because chemical species that occur in drinking water are known to be toxic at elevated concentrations, while their true hazard in lower concentrations remains unknown. In addition, very few studies have ever been conducted to determine the actual concentration of these chemical pollutants in human blood and urine.

The Centers for Disease Control (CDC) reported the results of a 1999 study [32] that tested for the presence of 27 chemicals in the blood and urine of 5000 individuals throughout the United States (see Exhibit 1.6). Future CDC studies will expand this list of chemicals to 100[29]. The results of the initial study showed that elevated concentrations of mercury and phthalates (from plastics) were found in the human body and especially in women and children. In addition, a Public Broadcasting System report on chemical industry trade secrets by Bill Moyers in March 2001 described the presence of industrial chemicals in human blood. A blood sample taken from Mr. Moyers was analyzed by the Mt. Sinai School of Medicine for 150 industrial chemicals and 84 chemicals were found to be present.

These reports clearly indicate that industrial chemicals permeate our bodies. The potential risk to human health is obvious. The only questions that remain are how many man-made chemicals are actually in the human body, what is the range of concentrations for each chemical and what are the health impacts? Such information will take a very long time to accu-

---

[27]Drinking water may be analyzed for regulated compounds but may not be analyzed at a levels below concentrations that are deemed "safe" and, thus, their concentrations are not reported.

[28]As we accept other risks in today's environment. For example, living in geographic regions that have floods, earthquakes, tornadoes and lighting; or participating in activities such as driving a car or flying in an airplane. These risks, however, have been well defined. It is much more practical to dislocate from an area with a known potential for natural disasters or not fly than to cease drinking water or breathing air.

[29]This number of compounds is insignificant to the number of chemicals manufactured and used in the United States.

Exhibit 1.5   Standards in a Vacuum—Mixed Messages

In California, which has long been in the vanguard of the environmental movement, a quiet debate is raging over what makes a standard and who should be responsible for setting them. Two different agencies, the Office of Environmental Health and Hazard Assessment (OEHHA) and the Department of Health Services Division of Drinking Water and Environmental Management (DDWEM), are both responsible for providing the public with maximum contaminant levels (MCLs) and public health goals (PHGs) for chemicals in drinking water. Why have multiple agencies create standards?

The DDWEM is the primary agency in the state responsible for setting and enforcing drinking water standards (MCLs). The mission statement for DDWEM states that the agency is responsible "for promoting and maintaining a physical, chemical, and biological environment which contributes positively to health, prevents illness, and assures protection of the public."

MCLs are enforceable regulatory standards under the California Safe Drinking Water Act of 1996 (Health and Safety Code Section §116365, and must be met by all public drinking water systems to which they apply. According to the California Health and Safety Code §116365(a) the Department of Health Services is to establish a contaminant's MCL "at a level as close as is technically and economically feasible to its public health goal (PHG), placing primary emphasis on the protection of public

mulate. This information is critical to the use and application of standard-based pollution control. Yet given the fact that toxicity studies can take many years to complete for each chemical, analytical methods need to be validated and calibrated for each chemical, and recognizing that there are thousands of chemicals in the environment, this is a seemly impossible task. Without this information, the public will always have to live with some unspecified level of risk.

POPULATION, POLLUTION, RISK AND PRECAUTION

The fact that the environment is damaged by man's activities is a natural consequence of our existence. This knowledge is as fundamental as knowing the properties of water. Most individuals intuitively understand that

(Exhibit 1.5, continued)
health (emphasis added)." DHS is therefore allowed to consider the technical feasibility of removing contaminants and the cost for removing contaminants (e.g., the very public debate about arsenic standards and the apparent reversal of the Federal government of what is an acceptable standard). To date, the DHS has established primary MCLs for 78 chemicals and 6 radioactive contaminants.

In a twist of legislative genius, OEHHA as the agency whose mission is "to protect and enhance public health and the environment by an objective scientific evaluation of risks posed by hazardous substance," was given the added responsibility of defining what is a risk free level for chemicals in drinking water. The same Health and Safety Code that requires DHS to establish a contaminant's MCL requires that OEHHA adopt PHGs based exclusively on public health considerations. PHGs are to be considered by DHS when establishing drinking water standards but are not enforceable standards. They are not supposed to "impose a regulatory burden on public water systems."

In a classic case of one hand not knowing what the other is doing, these two agencies have created conflicting standards. While it may have been envisioned that DHS could reach out to the toxicological experts at OEHHA for advise on the health implications of chemicals in the drinking water supply, the legislature has created a Catch 22 for the agencies involved. While DHS may create standards, they must seek OEHHA's input even though OEHHA's advice may be ignored when it is technologically infeasible (i.e., too expensive).

man's use and abuse of the earth's resources results in various impacts (i.e., dirty water, airborne odors and damaged landscapes, etc.). Most individuals also understand that as the human population increases, the damage to the environment increases.

Paul Ehrlich's prophecy of environmental doom from increasing population as presented in his 1968 book, *Population Bomb*, is even more relevant today. Population in the United States continues to increase as the result of both natural birth/death rates and unchecked immigration [33]. Clearly, as population and the gross domestic product of the United States increase, pollution increases. On the other hand, man's expanding use of natural resources coupled with technological advancements should not be

Exhibit 1.6   The CDC Report

The CDC's 2001 *National Report on Human Exposure to Environmental Chemicals* found that virtually all humans have some "background" level of industrial chemicals in their bodies. The report states that "...the measurement of an environmental chemical(s) in a person's blood or urine does not by itself mean that the chemical causes disease....For most of the other environmental chemicals [i.e., chemicals other than lead], we need more research to determine whether levels measured in the Report are of health concern."

The CDC knows that industrial chemicals can be found in human blood and urine, but does not know their impact on human health. To determine this impact will take decades of research. The research proposed by the CDC would appear to be an impossible task given the number of industrial chemicals used in the United States. As it is, the CDC report addressed only the following chemicals:

  • Metals: lead, mercury, cadmium, cobalt, antimony, barium, beryllium, cesium, molybdenum, platinum, thallium, tungsten and uranium.
  • Organophosphate pesticides (by measuring six common metabolites that are representative of the following pesticides): chlorpyrifos, diazinon, fenthion, malathion, parathion, disulfoton, phosmet, phorate, temephos, and methyl parathion.
  • Phthalate metabolites (phthalates are used in soap, shampoo, hair spray, nail polish and plastics) : mono-ethyl phthalate, mono-butyl phthalate, mone-2-ethylhexyl phthalate, mono- cyclohexyl phthalate, mono-n-octyl phthalate, mono-isononyl phthalate, and mono-benzyl phthalate.

Once again, this list of chemicals differs from those found in the USEPA Primary Drinking Water Quality Standards shown in Appendix A, as well as the chemicals listed in Exhibit 1.4. Clearly, the current number of pollutants present in water exceed the government's ability to even define the problem, no less offer a solution.

viewed solely as creating harmful impacts. In contrast to the destructive aspects of expansion, there has been an ever increasing production of food as well as significant advancement in medical science. As a result, humans on average (at least in the developed nations) are living longer and better. The dichotomy presented by these facts will suggest to some that the continuing and expanding modification of the environment has little or no apparent affect on human health. Such a simplistic conclusion, however, does not take into account the diversity within the human species and habitat.

Because humans are living longer, we are also exposed to both acute[30] and chronic[31] levels of pollutants for longer periods of time. The ultimate effect of this exposure is not scientifically known but must be a concern to a society that has come to anticipate a prolonged life expectancy. This concern is more than justified since medical science has clearly determined that young individuals with varying degrees of sensitivity to illness and the elderly with their own degrees of sensitivity are generally more susceptible to health problems. In other words, humans on average may be living longer, but the impact of chemical pollution on those who live on or near the edge of the bell-shaped curve is unknown.

The ultimate effect on humans exposed to a specific chemical pollutant or combination of pollutants is not scientifically known. This lack of knowledge is addressed by scientists as a "calculated risk." When calculating a risk, one must rely upon some degree of valid information to serve as a basis for the calculation and an estimate of the confidence in the resulting prediction. When considering the extremely limited amount of data on the exposure of humans to a specific toxic chemical or mixture of chemicals, there is currently no basis to assume that risk calculations can provide safeguards to the community. For example, a study by the United States Geological Survey [34] on pollution of the nation's water resources reports that, "the pervasive uncertainty of extrapolating results from laboratory animals to humans, estimating the risk associated with long-term consumption of drinking water that contains pesticides, even at levels below current regulatory standards, is speculative." In other words, the major problem with attempting to calculate the risk to humans is the lack of any realistic scientific database underlying the validity of the exercise.

---

[30]This is usually a one-time exposure to elevated levels of a pollutant. For example, an individual walking down a residential street inhales a high concentration of a pesticide that was just sprayed by a city or county agency to control mosquitoes.

[31] This is generally a fairly continuous exposure to low levels of a pollutant. For example, the ingestion of low concentrations of chemicals that are not removed from drinking water resources.

Regardless of these issues, the USEPA still relies upon the risk assessment process to determine health risks to humans from specific chemicals in drinking water.

A seminal event in public health and water quality occurred in 1849 when John Snow, a founding member of the London Epidemiological Society, set out to discover the source of a large waterborne outbreak of cholera that had occurred in London. By studying the distribution of cases, he was able to determine that the cholera was most likely coming from a public well that was being polluted with human and animal waste. To control the epidemic, he simply took the handle off the pump. Thirty-four years later, cholera vibrio, the source of the epidemic, was identified. Snow was able to make the association between disease and the sources—and end the epidemic—without knowing all of the details.

Today with multiple sources of pollution, a massive number of man-made chemicals in the environment, and the development of modern drinking water and waster water treatment systems, the *clear cause and effect relationship* between pollutants, water quality, and public health has been obscured. Thus, the ability to implement public health policy, and in particular water quality policy, to protect the public health from exposure to toxic compounds is confounded by a fundamental lack of human exposure data. No simple method exists for predicting the true impact of consuming low levels of man-made chemicals in drinking water. As a result, this problem cannot be solved by simply removing the "pump handle". Unlike John Snow, we now must spend millions of dollars and years of research deciding on the appropriate mode of action for ensuring sufficient water quality to protect the public health.

Because no real chemical exposure database exists for humans, environmental policy makers currently rely on an iterative process of risk management, risk communication and risk assessments as a means of addressing the unknown consequences of consuming chemically polluted drinking water. To understand the purpose and limitations of this iterative process, it is important to understand the basic components of risk management, risk communication and risk assessments.

Risk management is the process of weighing policy alternatives in light of the results of a risk assessment and implementing appropriate control options. The stated goal of risk management is to protect the public health by controlling risks as effectively as possible through the selection and implementation of appropriate measures. Risk communication is the exchange of information and opinion on risk among risk assessors, risk managers, and other interested parties, including the general public.

Risk assessment is a structured process for determining the potential risks associated with any type of hazard—biological, chemical, or physical.

The Risk Assessment Process

Risk assessment has been defined [35] as "a systematic process for making estimates of all the significant risk factors that prevail over an entire range of failure modes and/or exposure scenarios due to the presence of some type of hazard. It is a qualitative or quantitative evaluation of consequences arising from some initiating hazard(s) that could lead to specific forms of system response(s), outcome(s), exposure(s), and consequence(s)." The risk assessment process should be the mechanism by which the best available scientific knowledge is used to establish case-specific responses that will ensure defensible decisions for managing hazardous situations in a cost-effective manner. However, this approach begs the question - is this the best mechanism?

Risk assessments guidelines were originally published in the National Academy of Sciences "Redbook," which suggested that risk assessments should contain some or all of the following four steps [36]:

• Hazard Identification;
• Toxicity Assessment;
• Exposure Assessment; and
• Risk Characterization.

Hazard identification: This process concerns the discussion of the toxicological properties of a particular chemical. It is a qualitative evaluation that examines the applicable biological and chemical information to determine whether exposure to a chemical may pose a hazard or increase the incidence of a health condition or effect (e.g., cancer, birth defects, etc.) [37]. Human health effect studies are preferred over animal studies because of interspecies variation in response to chemical exposure. However, adequate human studies are generally not available for each chemical of concern. When this occurs, the results of animal studies are used to estimate the potential for human health effects from exposure to the same chemical.

For mixtures of chemicals, the toxicity may be evaluated in one of two ways. The first is to evaluate the mixture as a whole. This method is preferable, when adequate human and animal studies have been conducted, as it may account for various effects of chemical interaction, including antagonistic, synergistic and potentiative effects, resulting from one or more of the chemicals. Antagonists are chemicals that when combined diminish each other's effects (e.g., 2 + 4 = 3). Synergists are chemicals that have the same toxicity and when combined produce a greater effect than their additive effect (e.g., 1 + 1 = 20). Potentiation occurs when a chemical that does not produce a specific toxicity increases the toxicity of another chemical (e.g., 2 + 0 = 10). It is important to note that this type of information is usually developed from actual toxicity studies. The second method uses models to evaluate the toxicities of individual chemicals in a mixture when there is not adequate information to evaluate the mixture as a whole. In these cases, indicator chemicals (i.e., usually the most toxic or highest concentration in a mixture) are used to estimate the potential risk.

Toxicity Assessment: A toxicity assessment is the process of evaluating whether the possibility exists for an increase in the incidence of an adverse health effect (e.g., cancer, birth defect, etc.) due to human exposure to a substance. The assessment identifies the relationship between the dose of a substance and the likelihood of an adverse effect in the exposed population [38]. Two methods of dose-response analysis are widely used to estimate effects in humans to low exposure levels of chemicals.

The first is mathematical modeling of the dose-response relationship. The approach is often used to characterize the relationship between the dose of a defined carcinogenic chemical and incidence of cancer. Model outputs yield a cancer slope factor that represents an estimate of the ". . . largest possible linear slope (within the 95% confidence limit) at low extrapolated doses that is consistent with the data."[39]. A mathematical model, such as the non-physiologically based Linearized Multi-Stage model, is used in conjunction with experimental data (when available) for this purpose. This model assumes that there is some risk associated with any dose of the chemical, even one molecule (i.e., the One-Hit Theory).

Chemicals that exhibit carcinogenicity are generally considered to have no exposure threshold (i.e., exposure to any amount of the chemical would result in some risk of cancer). Most modeling for quantitatively estimating the carcinogenic nature of chemicals at low doses to which people are exposed under environmental conditions is based on studies regarding human exposure to radiation [40]. While this assumption may be appropriate for radiation, many members of the scientific community believe that this model may not be suitable for all chemical carcinogens. Radiation is known to be genotoxic (i.e., it reacts directly with DNA) and an initiator of cancer. As a result, the dose is linearly related to the amount of radiation received at the target organ. Upon closer review, this approach appears somewhat ludicrous. Since there is no "threshold" than one has to question the validity of a model that attempts to propose a method by which one can establish a "safe threshold level." Nevertheless, this is the approach the USEPA uses to quantify risk which, by definition, is unquantifiable.

Chemical carcinogens fall into at least three major categories [41]: cytotoxicants (i.e., chemicals toxic to cells), initiators, and promoters (i.e., chemicals that promote the growth of cancer cells).The USEPA uses the Linearized Multi-Stage low-dose extrapolation model as the basis for estimating chemical-specific cancer risk at low doses. This model is recognized as a conservative approach to ensure the potential risk is not underestimated. Cancer slope factors derived from the Linearized Multi-Stage model are indices of carcinogenicity and are used in performing quantitative calculations to estimate carcinogenic risk.

Although there are some data on human exposures, most available information about dose-response relationships are based on data collected from animal studies and theoretical perceptions about what might occur in humans. The nature and strength of the evidence of the causation of cancer in an important aspect of the evaluation.[36]

Carcinogenic classifications developed by the USEPA's Cancer Assessment Group classify candidate chemicals into one of the following groups, according to the weight of evidence for and against carcinogenicity from animal and epidemiological studies [37]:

- Group A - Human Carcinogen (sufficient evidence of carcinogenicity in humans)
- Group B - Probable human carcinogen
- Group B1 - Limited evidence of carcinogenicity in humans
- Group B2 - Sufficient evidence of carcinogenicity in animals with inadequate evidence in humans
- Group C - Possible human carcinogen (limited evidence of carcinogenicity in animals; absence of human data)
- Group D - Not classifiable as to human carcinogenicity
- Group E - Evidence of non-carcinogenicity for humans (no evidence of carcinogenicity in adequate studies)

The second method of assessing the dose-response relationship is through the safety factor approach. This method is used to describe the relationship between the dose and the effects of defined non-carcinogenic chemical. The induced effect is assumed to have a threshold below which adverse health effects would not be seen. Exposure levels that do not result in adverse health effects in animals are extrapolated to human exposures using safety factors. According to USEPA [42], a reference dose (RfD) or reference concentration (RfC) is a provisional estimate (with uncertainty spanning perhaps an order of magnitude) of the daily exposure to the human population (including sensitive subgroups) that is likely to be have no appreciable risk or deleterious effects during a portion of the lifetime, in the case of subchronic RfC or RfD, or during a lifetime, in the case of a chronic RfC or RfD.

Chemicals that exhibit adverse effects other than cancer or mutation-based developmental effects are believed to have a threshold (i.e., a dose below which no adverse health effect is expected occur). When extrapolating animal data to identify safe levels of human exposure, most researchers have focused on the use of a safety factor or uncertainty factor. The magnitude of the safety factor is, in turn, dependent on a number of quantitative and qualitative determinations based on the type, duration, and results of the animal research study. The concept of a threshold event is based on the assumption that there is a dose below which there is no effect. Health criteria levels are usually estimated from the no-observed adverse effect level (NOAEL) or the lowest observed adverse effect level (LOAEL) determined in chronic animal studies. The NOAEL is defined as the highest dose at which no adverse effects occur. The LOAEL is the lowest dose at which adverse effects begin to appear. NOEAL and LOAEL derived from animal studies are used by USEPA to establish oral and inhalation reference doses (RfDs) and reference concentrations (RfCs) for human exposures. The USEPA has used these approaches in establishing exposure route-specific reference doses (RfDs) for non-carcinogenic chemicals. A RfD is a daily dose level to which humans may be exposed throughout their lifetimes with no adverse health effect expected. A RfC is the concentration of a chemical in air to which humans may be exposed throughout their lifetimes with no adverse health effect expected.

The highest degree of uncertainty identified with most risk assessments is normally associated with the extrapolation of results obtained from animals tested at high doses to those results that would be anticipated at low doses, which humans are more likely to encounter in the environment.

Exposure Assessment: An exposure assessment, as defined by the National Academy of Science [36], is the process of measuring or estimating the intensity, frequency and duration of human exposure to an agent in the environment. The quantitative assessment of exposure, based on the chemical concentrations and the degree of absorption of each chemical, provides the basis for estimating chemical uptake (dose) and associated health risks.[37, 43].

Potential exposure to chemicals in the environment is directly proportional to concentrations of the chemicals in environmental media (e.g., water, air, etc.) and the characteristics of exposure (e.g., frequency and duration). The characteristics of exposure are estimated using various exposure parameters. The concentrations of chemicals at specific exposure points will vary over space and time. However, a single estimate of an exposure point concentration may be used for risk assessment. This single value must be representative of the likely or average concentration to which a person would be exposed over the duration of the exposure. However, to be conservative and health-protective, point estimates based upon concentrations greater than the average or most likely estimate of exposure are often used to avoid underestimating potential exposure.

Risk Characterization: Risk characterization is the description of the nature and magnitude of potential health risk, including attendant uncertainty. Risk characterization integrates the results of the exposure assessment and the toxicity assessment to estimate potential carcinogenic risks and non-carcinogenic health effects associated with exposure to chemicals. This integration provides quantitative estimates of either cancer risk or non-cancer hazard indices that are compared to standards of acceptable risk.

Importance of the Risk Assessment Process

The overall purpose of a risk assessment is to provide, in as far as is feasible, sufficient information to risk managers to allow the best decision to be made concerning a potentially hazardous problem [35]. Whyte and Burton [44] defined the major objective of risk assessment as the development of risk management decisions that are more systematic, more comprehensive, more accountable and more self-aware of what is involved than has often been the case in the past. Tasks performed during a risk assessment are intended to help answer the questions "How safe is enough?" or "How clean is clean enough?" The risk assessment process is supposed to be a transparent method for characterizing the nature and likelihood of potential harm to the public. It is also supposed to help define the uncertainties and provide some level of comfort with the inferences that are made. And, finally, it is supposed to point out data gaps that can help prioritize research needs.

Most major federal environmental and food safety laws require the use of risk analysis to determine what constitutes a safe level of chemical exposure. Those laws include the Clean Air Act, the Federal Water Pollution Control Act of 1972, Comprehensive Environmental Response, Compensation, and Liability Act of 1980, Federal Insecticide Fungicide, and Rodenticide Act, Hazardous Materials Transportation Act, Occupational Safety and Health Act, Resources Conservation and Recovery Act, Superfund Amendments and Reauthorization Act of 1986, the Safe Drinking Water Act, Toxic Substances Control Act, and the Food, Drug, and Cosmetic Act.

In addition to its widespread use in the promulgation of drinking water quality standards, risk assessments are used to set standards for air quality, soil cleanup goals, and to protect food quality [45]. When the Food and Drug Administration set out to revise the Recommended Dietary Allowances several years ago it was concerned about how to define a tolerable upper level for the intake of nutrients. This question was of great importance to nutritionists at the Food and Drug Administration, since the use of dietary supplements, in conjunction with the practice of food fortification, may provide high levels of nutrients that might put consumers at risk. So whereas regulations for water are focused on "how little is safe enough," the standards regarding nutrients in foods often focus on "how much is safe enough." Another example of the value of risk assessment in food safety is the recently completed risk assessment by United States Department of Agriculture, Food and Drug Administration and the Center for Disease Control scientists on *Salmonella Enteritidis* (*SE*) in eggs and egg products. This was the first quantitative farm-to-table microbial risk assessment and it is expected to serve as a prototype for future risk assessments. According to Food and Drug Administration, data from the Center for Disease Control indicated that *SE* is one of the most commonly reported causes of bacterial food-borne illness in the United States and has been increasing since 1976. Data from the risk assessment indicated that consumption of contaminated eggs results in an average of 661,633 human illnesses per year.

During the *SE* risk assessment, the Food and Drug Administration created a farm-to-table model that it can use to determine the effects of specific interventions on the incidence of illness. As part of this risk assessment, the team evaluated a number of possible interventions on the expected number of human illnesses, including shell egg cooling, diverting eggs from flocks with a high prevalence of *SE*-positive hens to breaker plants for pasteurization, and reducing the prevalence of *SE* -positive flocks.

Risk Assessment Determinations

Risk assessments determine the probability that an adverse health effect may occur in a population exposed to a toxic agent. Risk assessments cannot determine whether any one individual will become ill after an exposure to an agent. Therefore, a risk assessment is a basic predictive device for risk managers to estimate potential accidental exposures, establish food tolerances, set environmental pollution and cleanup levels, define allowable workplace exposures, and evaluate the risk of chemical pollution from uncontrolled waste sites. In order that risk assessments can be used for these purposes, target risk levels need to be defined.

Target risk levels that represent the tolerable limits to danger that society is prepared to accept as a consequence of potential benefits that could accrue. These levels may be represented by the *de minimis* or "acceptable" risk levels [35]. Risk is *de minimis* if the incremental risk produced by the activity is sufficiently small so that there is no incentive to modify the activity [46]. Various demarcations of acceptable risk have been established by regulatory agencies. Cancer risks in excess of 1 x 10-5 (1 in 100,000) per chemical have been deemed unacceptable pursuant to the California Safe Drinking Water and Toxic Enforcement Act of 1986, otherwise known as Proposition 65 (California Health and Safety Code Sections 25249,5 *et seq.*; 22 California Code of Regulations Section 12703(b)). The USEPA generally deems health risks to be significant if cancer risk exceeds the USEPA acceptable risk range of 1 x 10-6 to 1 x 10-4 (1 in 1,000,000 to 1 in 10,000) and/or the hazard index is greater than 1 (40 Code of Federal Regulations part 300.430(e)(2)(I)(A)(2).

Because the USEPA and state environmental agencies consider a risk assessment as an important predictive tool, drinking water standards are strongly tied to the risk assessment process. This does not mean, however, that the process is necessarily protective of the public health. This failing is fundamentally associated with a sequential process that magnifies any error or uncertainties in the basic assumptions of the predictive models. Although the government attempts to "build in a safety factor" in setting standards, the truth remains an open issue—we really do not know the effect of a chemical, even at the lowest levels.

Validity of the Risk Assessment Process

As with any assessment process, the quality and quantity of the data collected for the assessment will affect the validity of the assessment. The validity of any risk assessment is dependent upon the uncertainties inherent with each step in the risk assessment process. For example, the uncertainties associated with the area of *hazard identification* include [45]:

- The relative weights given to studies with differing results. For example, should positive results outweigh negative results if the studies that yield them are comparable?
- The relative weights given to results of different types of epidemiological studies. Are the results of a prospective study more valid than those of a case-control study?
- hould experimental-animal data be used when the exposure routes in experimental animals and humans are different?
- How much weight should be placed on the results of various short-term tests?

Uncertainties associated with the area of *toxicity assessment* include [45]:

- The dose-response models used to extrapolate from observed doses to relevant doses.
- How should different temporal exposure patterns in the study population and in the population for which risk estimates are required be accounted for?
- How should physiologic characteristics be factored into the dose-response relation?
- Should dose-response relations be extrapolated according to best estimates or according to upper confidence limits?
- What factors should be used for interspecies conversion of dose from animals to humans?

Uncertainties associated with the area of *exposure assessment* include [45]:

- Should one extrapolate exposure measurements from a small segment of a population to the entire population?
- Should dietary habits and other variations in lifestyle, hobbies and other human activity patterns be taken into account?
- Should point estimates or a distribution be used?
- How should exposures of special risk groups, such as pregnant women and young children, be estimated?

Uncertainties associated with the area of *risk characterization* include [45]:

• Statistical uncertainties in estimating the extent of health effects. How are these uncertainties to be computed and presented?
• Which population groups should be the primary targets for protection and which provide the most meaningful expression of the health risk?

Given this extensive list of uncertainties that are associated with the risk assessment process, it is no wonder that the validity of the process is suspect. However, it may well be that this is the best approach that science can furnish. But one must question whether mice and rats are really people?

Frustration over not having this type of chemical specific data for human exposure prompted researchers at the Loma Linda Medical Center in California to begin a study designed to determine if perchlorate in drinking water interferes with thyroid glands [47]. In this study, volunteers were paid $1,000 each to take pills containing perchlorate. Because this study is unique and privately funded, it is highly unlikely that the federal government will sponsor similar studies for other chemicals. This reality virtually assures that valid human exposure/response data will not exist for a substantial number of chemicals that currently pollute our drinking water resources. Without this type of information, drinking water standards cannot be shown to protect human health.

This conclusion is supported by an article in the Journal of the American Water Works Association [48]. This article quotes William Ruckelshaus, a former two-time administrator of the USEPA, as saying that "USEPA's laws often assume, indeed demand, a certainty of protection greater than science can provide given the current state of knowledge. The public thinks we know all the bad pollutants, precisely what adverse health or environmental effects they cause, how to measure them exactly, and control them absolutely. Of course, the public, and sometimes the laws are wrong." The article's author concludes that *"He hit the nail right on the head—we expect too much of science and government when it comes to defining what 'safe' is when dealing with trace amounts of chemicals."*

This uncertainty, however, can also be used to argue the opposite opinion. For example, Bjørn Lomborg contends in *The Skeptical Environmentalist* [49] that although "we possess only extremely limited knowledge from studies involving humans, and by far the majority of our evaluations of carcinogenic pesticides are based on laboratory experiments on animals," these studies show that the risk is negligible. Lomborg estimates that the annual cancer mortality due to pesticides in the United States is probably close to 20 out of 560,000 and that this is an acceptable risk given the benefits of pesticides to agricultural production and the control of disease. Yet, his conclusions would be more valid if the toxicity studies were based on human data. Again, are mice and rats people? Considering that researchers lack both human toxicity data and knowledge of the effects of consuming a low level mixture of chemical pollutants in drinking water as well as the additive chemical exposures in food and air, it would seem prudent to limit, to the maximum possible, our exposure to chemicals in drinking water.

When considering all of the issues associated with estimating risk and the policy conclusions that are based on such risk predictions, what level of trust should the public have in the risk assessment process? Unfortunately, the public's trust has been damaged in the past by the miscommunication of risk information–and such communication is as important as the actual quantification of the risk itself. Two examples of the miscommunication are the "confusion" associated with Alar and Dioxin.

In the case of Alar, the Natural Resources Defense Council mounted a major public relations campaign in 1989 designed to force USEPA to speed up pesticide regulation [50]. National coverage of the Alar controversy included risk estimates for which little or no explanation was given. Differences of opinion on the cancer potential of Alar, how many apples were treated with Alar, and the number of apples eaten by children resulted in significantly different risk estimates from the USEPA and the Natural Resources Defense Council. The confusion left the public panicked and resulted in ten cities including New York, Atlanta, Chicago and Los Angeles banning apples and apple products in school lunches. The apple industry estimated that it lost more than $100 million in apple sales. And most importantly, Americans' faith in the safety of the nation's food supply was shaken. Today, most scientists agree that the "concern" over Alar was unwarranted. In the second example, a series of articles in 1993 in the *New York Times*, which questioned the many standards used by USEPA for regulating toxic chemicals, highlighted the dioxin controversy. The environmental reporter for the *Times*, Keith Schneider [51, 52], wrote that "new research indicates that dioxin may not be so dangerous after all." He noted that many scientists and public health specialists said that "billions of dollars are wasted each year in battling problems that are no longer considered especially dangerous, leaving little money for others that cause far more harm." Here again miscommunication has resulted in misinformation. Inadequate or misleading risk communication by the media or scientific community contributes to the public's fear of environmental risks, such as Alar and Dioxin, or more recently arsenic and perchlorate. While the media play an important role in influencing policy decisions and regulations, they also play an equally significant role in the framing the discussion. Reporters must realize that even if they cannot place important risk assessment information into stories because of space or time limitations, they must understand this information to ask the right questions of sources and be sure they cover all of the important points.

When the questionable validity of risk assessments is combined with misleading communication to the public of the true risks, it is understandable that the risk assessment process creates an air of uncertainty. Given this uncertainty, it is not unreasonable to invoke the "precaution principle" [53] that states "when an activity raises threats of harm to human health or the environment, precautionary measures should be taken even if some cause and effect relationships are not fully established scientifically." The application of this principle is especially relevant to this problem since many inorganic and synthetic organic compounds are known to be harmful to humans at elevated concentrations. However, the decision to use this principle can also have widespread social and economic repercussions.

Obviously, determining the extent to which "precautionary measures" are needed is the major problem in attempting to address any threat to human health. Although the application is complex, there are also solutions, some of which are also complex, that can meet this challenge. The "precaution principle" must be incorporated into the pollution policies of today, while reasonable solutions are still viable.

A POLICY OF POLLUTION

Historically, the "precautionary measures" for controlling chemical pollution that have been embraced in the past focused on the use of a region's natural resources to either "assimilate" or "dilute" pollution residues. For example, a waste could either be discharged into a stream with no limitations on the amount released (i.e., not regulated) or be below an established regulated level. In either case, pollution is a sanctioned policy. These historic approaches reflect governmental programs, which advocate the use of either nature's assimilative capacity[32] to control the spread of pollution or treatment to remove pollutants to a regulated concentration prior to its release. Neither approach addresses the real problem, which is that gov-

---

[32]The solution to pollution is dilution.

[33]The reason pollution is acceptable is because it is assumed that it is economically prohibitive to eliminate pollution.

ernmental policy is based on the presumption that some level of pollution is acceptable.[33]

The absurdity of current federal "standard" based policy is well illustrated by historical warnings of the pollution of surface and groundwater as far back as the middle of the 19th century [54]. Numerous examples of surface water pollution by unregulated chemicals have been reported since the beginning of the 20th century [55, 56]. Likewise, evidence of chemical pollution from unregulated compounds has also been reported for groundwater in the 1930s, 1940s and 1950s [57, 58, 59]. Decades before Rachael Carson[34] came on the scene, the common unregulated weed killer, 2,4-D, was found to have polluted wells in Montebello, California. Similarly, TCE, which until recently was an unregulated compound, was identified as a groundwater pollutant as far back as 1949. Concern over groundwater pollution in 1953 prompted the underground waste disposal task group of the American Water Works Association [60] to state that "the proper time to control underground pollution is before it occurs." The American Water Works Association was only one of many who advocated this approach. Despite the warnings in the early 1950s, pollution policies remained unchanged.

Approximately 50 years later, historic pollution from unregulated chemicals is now being remediated at a great expense. Furthermore, the USEPA [61] has now recognized the failure of currently available technology to cleanup groundwater (i.e., once polluted groundwater resources may never be fully restored). By neglecting the presence of unregulated pollutants in water resources, industries has been forced to expend billions of dollars for remedial actions that are not totally capable of eliminating the hazard. The net result is that our water resource will remain polluted for decades or centuries.

One of the obvious reasons for studying history is to avoid repeating past mistakes. By clinging to a policy of "standard based pollution," the federal government continues to repeat the mistakes of the past and at a substantially greater cost. Standard based pollution was demonstrated to have been a bad policy 50 years ago and it remains a bad policy today.

Because standard based pollution is our nation's policy of choice, a complicated and overlapping environmental control code has been put in place to enforce pollution standards. In the United States, there are 32 fed-

---

[34]Author of the 1962 book *Silent Spring*, which eventually resulted in governmental policies to ban the use of DDT.

[35]Which includes 200 sets of federal rules, regulations and laws.

eral executive agencies in 10 cabinet departments, including the Executive Office of the President, that are active within 25 separate programs[35] to manage and protect water quality [62]. All of these programs are linked to the concept of water quality standards. As a result of this linkage, the USEPA adopted a quantitative method (i.e., the risk assessment process) for assigning a level of hazard to specific chemical standards.

The risk assessment process is used to convince the American public that our drinking water is safe. These statistical models have been used for several decades by the USEPA to justify the release of specified levels of a chemical pollutant into our water resources while simultaneously assuring the public that regulated pollution is "protective of human health and the environment." Faith in this method, however, should be tempered by the following statement on the reliability of a risk assessment as [63] ". . .at best a dubious exercise. In most cases it involves assumptions built upon other assumptions, the effect comparable to a house built on sand." Reliability of these methods is also suspect because no human toxicity data is available for use in this calculation. As with all models, the axiom "garbage in, garbage out" truly applies to risk assessment methods.

Individual citizens should not have the "burden of proof" shifted to them, when their health is potentially or actually threatened. Why is it that the1958 Delaney Amendment to the Food, Drug, and Cosmetic Act specified that "no" chemical additive shown to cause cancer in humans or animals could be added to food or cosmetics? Because policy makers of the time recognized the hazard to public health. Yet current regulations, including the 1996 Food Quality Protection Act, allow toxic and carcinogenic compounds at "accepted levels" in food as long as it is not an "additive." The list of pesticides that are allowed in food at specified "tolerance" levels is given in Appendix E.

The fact that humans are exposed to a mixture of chemical pollutants in both food and water suggests that total chemical exposures must be considered when evaluating their potential impacts on human health. It is not the intent of this book to address the risk of chemicals in food or the combined risk in food and water, however, this issue is briefly discussed in Exhibit 1.7. This issue aside, when the pesticides listed in the Primary Drinking Water Standards in Appendix A are removed from the list of pesticides regulated in food (Appendix E), there are still 302 pesticides left. Thus, 302 pesticides that are considered toxic in food are not deemed a hazard in water.

There may be various reasons for this dichotomy. The USEPA may assume that many of these chemicals have limited production volume, dis-

Exhibit 1.7 Chemicals in Food and Water

If one combines the chemical exposure from both food and water, the level of exposure is obviously increased. However, is this increase a greater risk? According to Bjørn Lomborg in The Skeptical Environmentalist [36], the risk is increased but is primarily associated with the food we eat. "The average American consumes 295 pounds of fruit and 416 pounds of vegetables a year. A rough calculation shows that he therefore consumes about 24 mg of pesticide each year. Even if one drinks 2 liters of water a day for a whole year from a well with a pesticide concentration exactly at the EU limit (which would be a maximally pessimistic scenario), one would absorb about 300 times less pesticide from the water than from fruit and vegetables." Such a general comparison does not negate the need to limit chemical exposure in drinking water nor does it address site-specific exposure or the potential synergistic effects of non-pesticide chemical mixtures. However, this finding creates another problem.

If foods produced without the use of pesticides (i.e., organic) are assumed to be truly free from pesticide residues (i.e., no unintended air pollution of a crop or cross contamination during processing), then individuals who want to eliminate their exposure to man-made chemicals should only consume organic foods and drink pure water. Such a decision, however, has a social cost because people who cannot afford this choice will continue to be exposed to higher levels of chemical pollutants.

tribution or solubility so that they would not pose a significant water pollution risk to the American public. However, they may pose a significant risk to the local residents in those areas where these chemicals are used. General presumptions about the distribution and fate of these pesticides by the USEPA provides no assurance that these chemicals are not in our water resources and drinking water supplies[36]. This dichotomy highlights the basic failure of chemical standards to protect human health. Unless a chemical is defined as a hazard to the American public, it is considered "nontoxic" or an "acceptable risk" by omission.

That is why governmental policies must be changed to embrace and

[36]There are no monitoring data that would rule out the occurrence of these chemicals in the drinking water.
[37]This should also be true for food but is beyond the scope of this book.

enforce the basic principle of the Delaney Act so that "no" chemical pollutant is allowed in our drinking water[37]. This conclusion is clearly justified given the circumstances. So how did we arrive at our current condition? Not surprisingly, the answer is "economics."

Simply put, there has been and continues to be an informal agreement between municipalities, industry and government to further their respective economic objectives. It has always been the objective of municipalities to treat water resources within their local budgets, while industry strove to reduce the cost of pollution controls and maximize their bottom line. The objective in both cases was to spend the least amount of money necessary to meet pollution requirements. In a similar fashion, it has also been the objective of government to maximize political advancement by balancing public demands for a cleaner environment against policies that would dampen economic growth.

For example, in March 2001 President Bush's administration put on hold a planned reduction in the arsenic standard for drinking water from 50 ppb to 10 ppb. The administration then partially reversed itself and mandated that the 10 ppb standard to be met by no later than January 2006. The reason offered by the Bush administration for delaying implementation of the 10 ppb standard was because of the need to make sure that this new standard was valid and affordable. This justification was preposterous given that (1) epidemiologic data suggest that the standard should be zero [64] and (2) the fact that no standard is valid. Indeed, the government's maximum contaminant level goal is currently set at zero. The real–and unspoken–reason for the delay was its potential cost to the U.S. economy. It has been previously estimated [65] that the annual cost of meeting this new standard would range from $6,494 to $1,340,716 per year[38] for each public water system. For example, when the 10 ppb arsenic standard is put into effect[39], the following cities would have to upgrade their water treatment systems [64]: Albuquerque, New Mexico; Chino Hills, California; Lakewood, California; Landcaster, California; Midland, Texas; Moore, Oklahoma; Norman Oklahoma; Rio Rancho, New Mexico; Scottsdale, Arizona and Victoria, Texas. All in all, approximately 4,000 out of the 74,000 systems regulated by this new arsenic level will have to install additional treatment in order to comply with the new standard [66]. Another reason given for the delay was that a reduced arsenic[40] standard would

---

[38]The reason for this cost range is due to the size of the public water system. The larger the system the greater the cost.

[39]Bush administration accepted the 10 ppb arsenic standard in October 2001 (a seven month delay).

[40]Arsenic is a pollutant in acid mine drainage from coal and metal ores with sulfides.

impact water treatment costs of mining operations in the United States. The government has historically negotiated with both municipalities and industry to reach an agreement on standards. Such agreements were and continue to be structured to "protect human health and the environment[41]" while not placing prohibitively expensive treatment requirements on municipalities and industry. This negotiated approach toward defining pollution standards has been used by the federal government, municipalities and industry for every major environmental regulation that dealt with the control of pollutant discharges. Because these agreements cannot be demonstrated to actually "protect human health and the environment," they appear to protect only the economy. There are estimates that at least a half-million illness per year can be attributed to the microbial pollution of drinking water [67] and yet the USEPA has no data on the number of illnesses from chemical pollution.

According to *Webster's Collegiate Dictionary*, safe is defined as "freed from harm or risk." Safe drinking water, therefore, must be "freed from risk." Based on the extent of water pollution throughout the country and today's diffuse boundary between drinking water and wastewater, our nation's drinking water cannot be shown to be "free from risk." As a consequence, the time has come for the public to demand an end to the use of failed drinking water standards and focus on alternative methods to aggressively protect the public health.

CHANGING POLICY

With the recent completion of the human genome project, we have now defined the single most complex and smallest element of human species. By defining the genetic code, man is poised on the brink of a major human and environmental re-design. Mapping the human genetic code is a marvelous achievement. Yet, we know virtually nothing of the large and small scale environmental factors (e.g., chemical pollution, electromagnetic pollution, habitat, foods-diet, medicine, stress, global warming) that can influence the health of the human species as a whole. Regardless of our ignorance, we have already begun the release of another unknown environment factor into our environment by pursuing the genetic manipulation of plants and animals.

In spite of the advances in the biological sciences, we still have failed to define the chemical hazards faced by humans. This failure, which is summarized in Exhibit 1.8 (*see* pp. 48–49), is responsible for the perpetuation

---

[41]The litany of the USEPA.

of water quality standards that are continuously outdated. As a result, our nation's policy of using standard-based pollution control methods to protect our drinking water may impact our health to a point beyond our ability to implement appropriate corrective actions. In today's world, we live in a habitat with unknown chemical risks, while seemingly protected by only "statistical assurances of safety." It is not unreasonable to demand that drinking water contain the minimum concentration of chemical pollutants that can be achieved using the best available technologies. A technology-based drinking water quality policy is not only a human health necessity, but is achievable within realistic cost parameters without threatening the national economy. Thus, all drinking water supplied to the American consumer in every home and business should be free of chemical pollutants or as close to zero as technically possible. To demand less foreshadows a pollution epidemic that may explode beyond our ability to remedy it.

A TECHNOLOGY-BASED DRINKING WATER QUALITY POLICY

Because the fundamental economics of industrial development in the United States are supported by governmental programs that allow chemical pollution of our waters, there is very little chance that a zero pollution discharge policy will become a reality. Nor are there as yet any practical solutions for controlling pharmaceutical and widespread non-point source pollution. These conditions virtually guarantee that pollutants will always be in our drinking water. From an objective viewpoint, this means that the only way to achieve chemical-free drinking water is to remove chemical pollutants from drinking water to the lowest levels possible. This requirement should also be imposed on many brands of bottled water (i.e., they should be labeled as to their point of origin and purity).

Given the diversity of public and private drinking water supplies in this nation, the implementation of a technology-based drinking water quality policy will require (1) tailored solutions based on appropriate technology considering local circumstances[42] and (2) a national testing and research program to evaluate best available technologies for both municipal and small in-home water treatment systems. Upgrading municipal water treatment systems will not be cheap. The USEPA [68] has estimated that approximately $1 trillion would be required to just meet existing infrastructure and water quality objectives, let alone to complete an upgrade to

---

[42]At a minimum, all water systems should have advanced oxidation, membrane filtration and a compatible post treatment disinfectant regime.
[43]These same technologies would also remove biologic hazards as well.

Exhibit 1.8   Reasons Why Water Quality Standards Cannot Be Shown to Protect Human Health

Water quality standards should not be used to protect human health for the following reasons:

- The vast majority of water quality standards are not based on actual human exposure data to specific chemicals. Damage to human health has not and is unlikely to ever be calibrated to a specific chemical concentration in drinking water* nor can these standards be validated in the real world.
- It is currently impossible to collect human exposure data because the human environment is too complex and toxicity studies for the most part will not use humans in the place of laboratory animals.
- The time period between the introduction of a chemical into the environment and when it is recognized as being toxic to humans or animals often spans decades. This problem will only get worse as we fall further and further behind in our evaluation of these chemicals.
- The overwhelming number of chemicals that have already been introduced into the environment have not, as yet, even begun to be evaluated to see if they are a hazard to humans.
- The rate at which new chemicals are introduced into the environment is significantly greater than the rate at which they are studied and recognized as being toxic to animals.
- There is virtually no data on the toxicity of chemical mixtures on animals, let alone humans.
- Given the current pace of evaluating chemicals for inclusion to the Primary Drinking Water Standards, it is impossible for the USEPA to

*Volunteers in a medical study conducted by Loma Linda Medical Center in California were being paid $1,000 each to take pills containing perchlorate in experiments aimed at establishing a drinking water standard for perchlorate (Las Vegas Review-Journal, November 2000).

maximum pollution removal.

However, even with these apparent economic obstacles, there are both short-term and long-term solutions for upgrading municipal water treatment systems to deliver water with minimum chemical pollutants[43]. Until these technologies are implemented at the municipal level, appropriate and

(Exhibit 1.8, continued)

identify and regulate the current number of chemicals already in the environment, let alone the approximately 2,000 new chemicals introduced each year.

• No consensus exists within the various federal programs as to which chemicals pose a threat in food and which should be regulated in drinking water. Nor is there a justifiable scientific basis for selecting those chemicals for which drinking water standards are set. By their omission, unregulated chemicals are therefore defined as "nontoxic" or as being an "acceptable risk."

• The methods used to select chemicals for potential inclusion to the Primary Drinking Water Standards are flawed.

• Even if a chemical is identified as being toxic to humans, economic considerations may preclude its addition to the Primary Drinking Water Standards.

• When standards are exceeded, the offending pollutant is not required to be removed from the drinking water supply. Consumers are only warned not to drink the water. For many individuals, this may not be a viable option.

• When unregulated chemicals are identified in drinking water, the offending community water supply neither has to notify its customers of the chemicals presence nor remove it.

• Disinfection processes used in treating drinking water are known to add unregulated carcinogens to drinking water.

• Given the potentially large number of chemicals that can be found in drinking water, it is impossible to monitor for specific unsuspected compounds.

Given these facts, water quality standards should not be used as a guide to governmental policy that is supposed to "protect human health and the environment."

currently available point-of-use water treatment technologies can be installed in homes, workplace environments and at isolated recreational areas. Furthermore, for those concerned with potential terrorist acts, the implementation of point-of-use water treatment systems will also provide the ultimate defense against the intentional chemical or biological pollution of our drinking water resources.

A technology-based drinking water quality policy will actively protect the public health by removing both known regulated and unregulated chemical pollutants in our drinking water. The need for a technology-based drinking water quality policy is further illustrated by the fact that current standard-based policy is now open to lawsuits. A recent article in the Journal of the American Water Works Association states [69] that "of particular concern to water suppliers is the litigation potential linked to the expanding body of research indicating possible reproductive developmental risks from relatively short exposures by pregnant women to elevated levels of certain DBPs, which have been and continue to be regulated based on a potential cancer risk associated with chronic exposure." In this same article, the Deputy Director of Government Affairs for the American Water Works Association remarked that "compliance with a standard should protect a drinking water utility from liability for any damages that might be associated with a substance or contaminant in drinking water....The whole point of a standard is reducing risks to levels that are achievable and socially acceptable." If a technology-based program was in place, this concern would not even be an issue.

The facts demonstrate that *the threat is real*. The alternative, continued reliance upon drinking water standards to protect the public heath, will only perpetuate a false sense of security. Sadly, if federal, state and local governments cannot overcome the inertia of change, this policy may ultimately be settled by litigation.

REFERENCES

1. Lichtenstein, P. et al., "Environmental and Heritable Factors in the Causation of Cancer—Analyses of Cohorts of Twins from Swedan, Denmark, and Finland," *New England Journal of Medicine*, Volume 343, No. 2 (July 13, 2000).

2. National Research Council, *Setting Priorities for Drinking Water Contaminants*, National Academy Press, Washington DC (1999).

3. Colvig, Timothy, "Evidence issues: Getting Expert Opinions Past the Judicial Gatekeeper and into Evidence," In Sullivan, Agardy and Traub, *Practical Environmental Forensics*, John Wiley & Sons, New York (January 2001).

4. Goodell, Edwin B., "A Review of the Laws Forbidding Pollution of Inland Waters in the United States," U.S. Geological Survey, Water-Supply and Irrigation Paper No. 103 (1904).

5. U.S. Public Health Service, "Report of Advisory committee on Official Water Standards," *Public Health Reports*, Vol. 40, No.15 (April 10,1925).

6. National Research Council, *Identifying Future Drinking Water Contaminants*, National Academy Press, Washington DC (1999).

7. USEPA, "Providing Safe Drinking Water in America" Office of Enforcement and Compliance Assurance, Washington, DC, EPA 305-R-00-002 (April 2000).

8. Cohen, Brain and Richard Wiles, "Tough to Swallow, how Pesticide companies Profit from Poisoning America's Tap Water," *Environmental Working Group* (August 1997).

9. USEPA, "National Drinking Water Contaminant Occurrence Database," *Envirofacts Warehouse* (April 28, 2000).

10 Fisher, L.M., "Pollution kills fish," *Scientific American*, 160: 144–146 (1938).

11. Dugan, P.R., *Biochemical Ecology of Water Pollution*, Plenum/Rosetta, New York (1972).

12. Chang, L.W., *Toxicology of Metals*, CRC Lewis Publishers, Boca Raton, FL (1996).

13. Klaassen, C.D., *Casarett & Doull's Toxicology*, McGraw Hill, New York (1996).

14. Krebs, R.W., *The History and Use of Our Earth's Chemical Elements*, Greenwood Press, Westport, CT (1998).

15. Scharfenaker, Mark A., "Chromium VI: a review of recent developments," *American Water Works Association*, Vol. 93, No. 11 (November 2001).

16. Browning, E., "Toxicology of Industrial Organic Solvents," *Medical Research Council Industrial Health Research Board*, London: Her Majesty's Stationery Office (1953).

17. DeZuane, John, *Handbook of Drinking Water Quality*, Van Nostrand Reinhold, New York (1997).

18. Committee on Environmental and Natural Resources, "Interagency Assessment of Oxygenated Fuels," Executive Office of the President, National Science and Technology Council (June 1997).

19. Gullick, R. W. and M. W. LeChevallier, "Occurrence of MTBE in Drinking Water Sources," *Journal of the American Water Works Association*, Vol. 92, No. 1 (January 2000).

20. Renner, Rebecca, "Scotchgard Scotched," *Scientific American* (March, 2001).

21. USEPA, "Region 9 Perchlorate Update," United States Environ-

mental Protection Agency, Region 9, 75 Hawthorne Street, San Francisco, California (June 1999).

22. Roefer, Peggy, "Endocrine-Distrupting Chemicals in a Source Water," *Journal of the American Water Works Association*, Vol. 92, No. 8 (August 2000).

23. USEPA, Endocrine Disruptor Screening Program, Report to Congress (August 2000).

24. Trussell, Rhodes, "Endocrine Disruptors and the Water Industry," *Journal of the American Water Works Association* (February 2001).

25. Ternes, T. A., "Occurrence of drugs in German sewage treatment plants and rivers," *Water Research* 32 (11): 3245–3260 (1998).

26. Hun, Tara, "Water Quality—Studies Indicate Drugs in Water May Come from Effluent Discharges," *Water Environment & Technology* (July 1998).

27. Kolpin, Dana W., et al., "Pharmaceuticals, Hormones, and Other Organic Wastewater Contaminants in U.S. Streams, 1999–2000: A National Reconnaissance," *Environmental Science and Technology* (March, 2002).

28. Regush, Nicholas, "Questions on Protecting US Water Supplies," *Journal of the American Water Works Association* (February 2002).

29. National Cancer Institute, *Atlas of Cancer Mortality in the United States*, NIH Publication No. 99-4564, 2000.

30. Maxwell, Steve, "Ten Key Trends and Developments in the Water Industry," *Journal of the American Water Works Association*, Vol. 93, No. 4 (April, 2001).

31. Saunders, Robin G., Letters to the Editor on "What You Didn't Think You Wanted to Know About Recycled Wastewater," *Scientific American* (June 2001).

32. Centers for Disease Control, "National Report on Human Exposure to Environmental Chemicals," Press release dated 3-21-2001.

33. U.S. Bureau of the Census, Population Division (March 30, 2000).

34. Gilliom, Robert J., Jack E. Barbash, Dana W. Kolpin and Steven J. Larson, "Testing Water Quality of Pesticide Pollution," *Environmental Science and Technology*, Vol. 33, No. 7 (1999).

35. Asante-Duah, K., *Hazardous Waste Risk Assessment*, Lewis Publishers, Boca Raton, Florida (1993).

36. National Academy of Sciences, *Risk Assessment in the Federal Government: Managing the Process*, National Academy Press, Washington, D.C. (1983).

37. USEPA, "Risk Assessment Guidance for Superfund: Volume I, Human Health Evaluation Manual (Part A), Interim Final," Office of Solid

Waste and Emergency Response, EPA/540/1- 89/002 (December 1989).
38. Preuss, P.W. and A.M. Ehrlich, "The Environmental Protection Agency' Risk Assessment Guidelines," *J. Air Pollution Control Assn.* Vol. 37 (1987).
39. USEPA, Integrated Risk Information System (2002).
40. Paustenbach, D.J. (editor), *The Risk Assessment of Environmental and Human Health Hazards: A Textbook of Case Studies*, John Wiley and Sons Publishing, New York (1989).
41. Anderson, M.E., "Quantitative Risk Assessment and Industrial Hygiene," *American Industrial Hygiene Association Journal* (1988).
42. USEPA, Supplemental Guidance to RAGS: Calculating the Concentration Term, Office of Solid Waste and Emergency Response, EPA/9285/7/081 (1992).
43. California Environmental Protection Agency, Supplemental Guidance for Human Health Risk Assessments and Preliminary Endangerment Assessment (1994).
44. Whyte, A. V., and I. Burton (editors), *Environmental Risk Assessment, SCOPE Report 15*, John Wiley and Sons, New York (1980).
45. Woteki, C, 1998. "Nutrition, Food Safety, And Risk Assessment-A Policy-Maker's Viewpoint," Remarks prepared for delivery by Dr. Catherine Woteki, Under Secretary for Food Safety, before Purdue University, West Lafayette, Indiana (June 1998).
46. Whipple, *De Minimus Risk, Contemporary Issues in Risk Analysis*, Vol. 2, Plenum Press, New York (1987).
47. Rogers, Keith, "Drinking Water Study: Volunteers ingesting pollutant perchlorate," *Las Vegas Review-Journal* (November 28, 2000).
48. Hoffbuhr, Jack W., "Water Scape an Executive Perspective, Risky Acronyms," *Journal of the American Water Works Association* (April 2002).
49. Lomborb, Bjørn, *The Skeptical Environmentalist, Measuring the Real State of the World*, Cambridge University Press, United Kingdom (2002).
50. Friedman, Sharon, *The Media, Risk Assessment and Numbers: They Don't Add Up*, Risk: Health, Safety & Environment, Franklin Pierce Law Center (1996).
51. Keith Schneider, "New View Calls Environmental Policy Misguided," *The New York Times* (March 21, 1993).
52. Keith Schneider. "U.S. Backing Away from Saying Dioxin is a Deadly Peril," *The New York Times* (August 15, 1991).
53. Appell, David, "The New Uncertainty Principle," *Scientific*

American (January 2001).

54. Peckston, T., *A Practical Treatise on Gas Lighting*, Herbert: London (1841).

55. Wilson, H. & Calvert, H., *Trade Waste Waters: Their Nature and Disposal*, Lippincott: Philadelphia (1913).

56. Kraft, R., "Locating the Chemical Plant," *Chemical & Metallurgical Engineering*, 34 (11): 678 (1927).

57. Brown, W., "Industrial Pollution of Ground Waters," *Water Works Engineering*, 88 (4): 171 (1935)

58. Pickett, A., "Protection of Underground Water from Sewage and Industrial Wastes," *Sewage Works Journal*, 19 (3): 464 (1947)

59. Muehlberger, C., "Possible Hazards from Chemical Contamination in Water Supplies," *Journal of the American Water Works Association*, 42 (11): 1027 (1950)

60. American Water Works Association, "Findings and Recommendations on Underground Waste Disposal: Task Group Report," *Journal of the American Water Works Association*, 45 (19: 1295 (1953)

61. U.S. Environmental Protection Agency, Groundwater Cleanup: Overview of Operating Experience at 28 Sites, EPA 542-R-99-006, September 1999.

62. United States of America's submission to the 5th Session of the Commission on Sustainable Development (April 1997).

63. Brown, Michael, H., *The Toxic Cloud, The Poisoning of America's Air*, Harper & Row, New York, NY (1987).

64. Albert, Mark, "A Touch of Poison," *Scientific American* (June, 2001).

65. USEPA, Technical Fact Sheet: Final Rule for Arsenic in Drinking Water, EPA 815-F-00-019 (January, 2001).

66. Pontius, Frederick W., " Regulatory Compliance Planning to Ensure Water Supply Safety," *Journal of the American Water Works Association* (March 2002).

67. EPA, "Liquid Assets 2000: American's Water Resources—a Turning Point, Office of Water, Washington, D.C., EPA-840-B00-001 (May, 2000).

68. *Federal Water Review* (September–October 2000).

69. Scharfenaker, Mark A., "Reg Watch, Water Suppliers Carefully Watching Liability Suits," *Journal of the American Water Works Association* (April 2002).

Chapter 2

Pollution

"The most alarming of all man's assaults upon the environment is the contamination of air, earth, rivers, and sea with dangerous and even lethal materials."

— Rachel Carson, *Silent Spring*, 1962

The increasing complexity of chemical pollution is evolutionary. As society has become more technologically advanced, pollution has evolved from primarily hazardous biologic constituents to an ever increasing mixture of toxic man-made chemicals. In addition to the increasing complexity of pollution, the sources of pollution are now distributed throughout every region of the United States. In response to the expansion and complexity of industrial pollution, methods for treating pollution have also evolved. However, primarily due to cost, advanced treatment technologies have not been widely implemented and non-point sources of pollution remain virtually uncontrolled. As a result, a wide range of regulated and unregulated pollutants from a variety of sources are being discharged into our water resources. Because of the chemical complexity of pollution today, the environment simply cannot assimilate all of these toxic discharges. Thus, pollution of drinking water resources has become unavoidable and the safety of water resources can no longer be guaranteed.

THE EVOLUTION OF DOMESTIC AND CHEMICAL POLLUTION

The first serious waste disposal problem encountered by humans was the disposal of their own waste products. Because these wastes were biodegradable[1],

---

[1]Organic wastes could be degraded by microorganisms into simpler compounds (e.g., carbon dioxide, water, and less complex organic materials).

the spread of disease associated with these wastes posed the greatest threat. But as long as the population remained small and widely distributed, the disposal of human waste usually created little or no problem. With increasing population and the growth of cities, waste disposal became a greater nuisance as well as a major health problem.

In an urban environment, the solution to this dilemma was to discharge waste into sewers or existing storm drains and then out of the city itself. In most cases, these wastes ultimately discharge to nearby rivers and their downstream communities. The impact on rivers, which were unable to effectively dilute or biodegrade an ever increasing waste load, was described in an 1885 report [1] to the Boston Health Department as leaving surrounding communities "enveloped in an atmosphere of stench so strong as to arouse the sleeping, terrify the weak, and nauseate and exasperate everybody."

It rapidly became clear to many communities that the "solution to pollution" was not just "dilution." Although sewage discharged to a body of water can be decomposed by microorganisms and further diluted, the capacity of a body of water to handle sewage flows is not infinite. One answer to this problem was to put the sewage back onto the land, thereby reducing discharges to surface bodies of water. An 1896 water supply text [2] reported that the soil could be used to "purify" sewage but the soil would be "overtaxed" if too much waste was applied and the groundwater polluted[2]. Thus, at the turn of the 20th century, sewage was managed by either dilution and biodegradation in rivers, lakes and oceans or purified by the action of soil microorganisms. Scientists and engineers at the time referred to nature's ability to handle such waste as its "assimilative capacity." In other words, biological wastes could be assimilated with little or no nuisance as long as the nature's capacity to purify waste was not exceeded.

To ensure the environment would not be overtaxed, individual states passed laws to control pollution in the 1890s to early 1900s [3]. For example, a 1902 New Jersey statute contains the following language: "It shall be unlawful for any person, corporation, or municipality to build any sewer or drain or sewerage system from which it is designed that any sewage or other harmful and deleterious matter, solid or liquid, shall flow into any of the waters of this State so as to pollute or render impure said waters, except under such conditions as shall be approved by the State sewerage commission."

---

[2]Thus, transferring the pollution from surface water to groundwater.

These laws tried to address concerns about waste discharges by regulating the amount and type of wastewater that could be released to surface bodies of water. However, with increasing population and industrial development, pollution began to migrate across state lines and forced a basin-wide approach toward pollution control. For example, a 1907 report on stream pollution described pollution in the Potomac Valley [4]: "The prosperity of the industries of the Potomac Valley, with its attendant increase of population, is justly a cause of congratulation to the several States within which the basin lies. Yet this success brings responsibilities that can not be shirked, . . . Acts which may be viewed with indifference in a sparsely settled country become crimes in densely populated communities. No resource will be more seriously affected by changed conditions than water. . . . For, one by one, the sources of pure water which are not too expensive to utilize will be preempted, and then will come the time when the supplies that have been ruthlessly damaged must be purified. . . .The silver river threads are direct lines of communication between each individual and every other below him on the stream. The offenses that he commits against the water are paid for by his fellow countrymen in the basin, and the bill is larger or small according to the gravity of the transgressions."

The industries responsible for the pollution in the Potomac River Valley included mining, leather tanning, manufacture of textiles, manufactured gas from coal (commonly referred to as illuminating gas for gas lights) and the manufacture of whisky. Wastes discharged from these industries, with the exception of the manufacture of whisky, were significantly different than domestic sewage. These wastes contained non-biological pollutants, which could not be "purified."

For example, mining, tanning and textile processes released toxic metals (e.g., arsenic, cadmium, chromium, copper, lead, mercury, silver and zinc), while manufactured gas and textile industries released toxic organic coal tars and dyes to the environment. Both the trace metals and organic compounds released by these industries could not be substantially biodegraded. As a result, these wastes accumulated within the environment and became a continuous source of toxic materials that could leach into a water resource for many years. When non-biodegradable chemicals (i.e., persistent chemicals) are discharged, one cannot rely upon nature's assimilative capacity to repair this insult to the environment and, as a result, past chemical discharges continue to pollute our water resources.

In the early years of industrial development, persistent wastes were discharged without any regard for the environment. Today, industrial waste discharges are controlled to a much greater degree. However, discharge permits still rely upon nature to assimilate or reduce the level of pollution. As a result of this policy, thousands of complex chemicals are allowed into water resources because it is assumed that they can be assimilated by the environment. Yet, for the vast majority of these chemicals, their environmental fate is unknown. Most of these compounds are also unregulated. Thus, their occurrence in the environment generally goes undetected because monitoring of these chemicals in our water resources is not required. As long as these chemicals remain unregulated and unmonitored, the government appears to be more than happy to adopt the policy "out of sight, out of mind."

The Evolution of Wastewater Control and Treatment

The combination of municipal sewage and industrial waste discharges coupled with an expanding economy required that pollutant waste loads be reduced. Steps were taken to ensure that solid wastes were not be dumped into surface waters and that wastewater was treated to remove some fraction of its chemical and biological constituents prior to its discharge to surface waters. These actions gave rise to the development of centralized sewage and industrial treatment facilities.

Throughout the early 1900s to the 1940s, more and more states imposed stricter pollution control requirements on the discharge of both sewage and industrial wastes. These state regulations often limited the amount of chemicals and bacteria in sewage and industrial waste discharges so as to reduce the pollution load on receiving waters. Greater emphasis was also placed on "engineered" waste treatment prior to waste discharge to the environment. On a community-wide or region-wide basis, this emphasis led to the development of publically owned treatment works (POTW), contained the following basic treatment processes:

- Physical methods (e.g., screens and racks) were used to remove debris from wastewater.
- Treatment of the liquid portion of the waste stream was accomplished by coagulation and settling to remove suspended solids (sludge) followed by biological treatment using trickling filters or activated sludge units.
- The final effluent would then be chlorinated to destroy microorganisms such as coliform bacteria.
- The sludge resulting from this treatment would be placed in an anaerobic[3] digester where microbial decomposition and the resulting elevat-

---

[3]This is an oxygen deficient environment.

ed temperatures served to disinfect sludge. The sludge (commonly referred to as "biosolids") was then either land farmed (tilled into soil), placed in a landfill, dumped in the ocean or incinerated.

• The treated wastewater could, depending upon the geographic location, be either evaporated, discharged into an adjacent body of water, used for irrigation or injected into the groundwater.

A POTW receives primarily sanitary waste from homes, commercial business and industry. It also accepts commercial and industrial waste, if these wastes do not contain levels of toxic chemicals sufficient to kill the microorganism in the plant's liquid treatment systems and digesters (i.e., causing an "upset"). However, the actual chemical makeup of industrial discharges was often unknown to the operators of a POTW because state wastewater permits during this period rarely required monitoring for any chemical pollutants. However, operators had other ways to determine the presence of chemical pollutants in a waste discharge. For example, the business or industry could disclose the actual composition of their wastewater. However, they rarely did. More frequently, an unknown pollutant would only be identified when an industry discharged a enough of that chemical to a POTW to cause a treatment upset. As a result, the real composition of wastewater discharges was unknown as long as no upsets occurred.

Like industries, residential dischargers also disposed of an equally complex array of organic chemicals down their sinks, drains and toilets. Because of these sources, most POTWs received an extremely diverse mixture of waste that contained a wide variety of chemicals, including:

• Household cleaning chemicals, medicines, petroleum products, solvents, and poisons,
• Human excrement that contained pharmaceuticals and their byproducts,
• Chemicals from commercial/light industry that included process chemicals (e.g., degreasing solvents, ink, dyes, petroleum, paints and paint solvents, dry-cleaning solvents, poisons, metal plating chemicals, acids, etc.) used in everyday business operations, and
• Industrial discharges containing a wide spectrum of both organic and inorganic chemicals.

The standard methods of wastewater treatment described above have the ability to remove many of the organic and inorganic compounds found in raw effluent. However, POTWs must rely upon dischargers to limit the release of chemical pollutants. For example, with 41 states having issued fish-consumption advisories due to mercury pollution and new analytical methods allowing measurement of mercury down to the low part-per-trillion, municipal and industrial facilities that discharge mercury will not only have to modify their monitoring for mercury, but also face more stringent control of mercury in their wastewater [5].

However, unregulated pollutants are another issue. The one exception is the monitoring for estrogen in wastewater. Researchers at the University of West of England (Bristol) have developed a very sensitive method for testing for low levels of estrogen in wastewater effluent[4] and river water [6]. Although monitoring for estrogen and hormones is important in order to define the extent of any problem, it is unlikely that such monitoring will result in the installation of additional wastewater treatment. Unfortunately, a vast array of pharmaceuticals, chemicals from personal care products [7] and disinfection by-products (DBP)[5], other than estrogen, are present in treated wastewater for which neither monitoring requirements nor simple methods of monitoring are available. Because most pharmaceuticals occur in wastewater due to the public's use of medications, there is no easy way for regulatory agencies to impose discharge limits.

Given this problem, the only method available for protecting the environment from these discharges is to rely upon wastewater treatment to remove these pollutants. Doing so is not a simple task. For example, researchers at the University of California at Berkeley [8] predict that 90 percent of hormones can be removed by a municipal wastewater treatment plant, but some compounds like NDMA are not removed at all. Because there are so many unregulated chemicals discharged to POTWs and so few actual studies on the "pass through" of specific organic chemicals, the extent of the problem is truly unknown. In other words, we have just begun to define the problem of micropollutants in wastewater discharges. As a result, scientists recommend that additional treatment methods be evaluated and developed for treatment of such pollutants. More specifically, they recommend the use of ultraviolet (UV) light or ozone in place of chlorine/bromine disinfection to eliminate disinfection by-products such as NDMA and reverse osmosis to remove additional and unspecified chemi-

---

[4]The most important application of this method will be for the monitoring of estrogen in drinking water.

[5]For example, N-nitrosodimethylamine (NDMA) which has a California MCL of 0.002 ppb and is commonly found in treated wastewater and groundwater where treated wastewater was used to recharge aquifers [9].

cal pollutants. Without using these treatment methods, most of these unmonitored chemicals will continue to be discharged into the environment unnoticed.

In addition to POTW discharges, many industries throughout the early 1900s to the 1940s also discharged their waste directly to water resources. In virtually all cases, industries would discharge chemically polluted wastewater into a water resource during periods of high flow when dilution would mask any observable damage. However, some industries as part of their discharge permit requirements did test for chemical and biologic indicators of pollution in the receiving body of water after dilution had occurred. The standard test methods used are discussed in Exhibit 2.1. In the summer months when river flow was low (i.e., less dilution) or when a chemical pollutant even at low concentrations would create observable changes in receiving water quality, the responsible industry would have to either (1) build holding ponds so that wastewater could subsequently be released during periods of high flow or (2) provide treatment prior to discharge. A good example of how industry dealt with wastewater treatment during this time period is illustrated in articles published by the Dow Chemical Company [10,11].

By the early 1940s, the Dow Chemical Company at its Midland, Michigan facility manufactured various industrial chemicals, organic solvents, pharmaceuticals, aromatic organic compounds, insecticides and dyes[6]. Dow identified and focused on only one class of chemicals (phenols) that was recognized as causing a major problem in its wastewater discharge. According to Dow [12], "phenol is used as a standard in testing disinfectants and its germicidal action is utilized in many ways, for example, in the manufacture of germicidal and disinfecting paints and germicidal soaps, and is a preservative in leather, glue and adhesives industries." Thus, phenol is clearly toxic[7]. The 1946 United States Public Health Service Drinking Water Standards [13] state that "phenolic compounds should not exceed 0.001 ppm."

According to Dow [10], the manufacture of phenol from benzol and the production of salicylates added chemical wastes that could be objectionable in water in very "minute amounts." Because phenol concentrations in their waste streams were too large to release even during the periods of

[6]For a list chemicals manufactured by the Dow Chemical Company in 1938, see Appendix F. This list also provides an example of the types of chemicals that were potential sources of pollution over 60 years ago.
[7]The 1946 Manufacturing Chemists= Association, Chemical Safety Data Sheet for Phenol establishes that chronic poisoning by phenol may be fatal.

Exhibit 2.1: Monitoring Pollution

Prior to World War II and through the early 1960s, the major constituents found in domestic sewage were from households as opposed to industrial sources. Because these wastes were found to be relatively biodegradable, methods of monitoring pollution measured the effect of these biodegradable wastes on water quality. The standard methods used to monitor the effects of biodegradable wastes were the biological oxygen demand (BOD), suspended solids, floating solids., pH, coliform organisms and dissolved oxygen (DO). Because these methods were used to characterize the degree of pollution in receiving waters, these parameters were also used to address and establish effluent discharge standards. In other words, the receiving waters were assumed to be protected if the specified limits were not exceeded. For example, the following parameter limits were in general use:

BOD: The BOD is a measure of the amount of oxygen consumed as biodegradable material is aerobically decomposed. Thus, the greater the BOD, the less oxygen available in the water and the greater potential harm to fish and aquatic organisms. BOD limits are usually set at 10 to 20 ppm to ensure that Aoverload@ conditions, which would cause excessive oxygen depletion in the receiving body of water, do not occur.

Suspended Solids: Suspended solids were usually limited to 20 to 100 ppm. This limit was necessary to prevent either excessive sludge build-up in the receiving water and to ensure that sunlight is not prevented from pene-

dilution (i.e., high flow conditions in the Tittabawassee River), Dow built a treatment plant designed to remove phenols. Prior to this plant going on-line, the concentration of phenol in Dow's wastewater ranged from 305 to 458 parts-per-million (ppm). After treatment operations began, phenol concentrations in the treated wastewater dropped to 0.01 to 1.7 ppm. This reduction is a good example of just what can be accomplished through good engineering practice, when the incentive exists to address a problem, even in the early 1940s.

(Exhibit 2.1, continued)
trating the water environment (i.e., turbidity associated with suspended solids could effect the growth of marine flora).

Floating solids:	This limit was an esthetic standard that required the full removal of all floating solids.
pH:	In order to maintain a natural acid/base balance, pH was usually set between 5.5 and 8.5.
Coliforms:	Coliform limits were determined by "frequency of tests" with a statistical average and allowable maximums over time. Actual values were usually set as follows: the most probable number (MPN) of coliforms should not exceed more than one per milliliter in 50 percent of the one milliliter samples.
Dissolved Oxygen:	Dissolved oxygen was usually set at 5 ppm to maintain this level in the receiving waters.

In addition to these general standards, each region of the country often had additional requirements based on local conditions. For example, discharges in the Ohio River Basin usually had restrictions on chlorides. Other parameters such as metals (e.g., zinc, copper, arsenic, cadmium, etc.), sulfides, nitrates, phosphates and phenols were occasionally incorporated into state discharge permits. The concentration limits were usually determined by the state and local governments since there were no national standards.

These same general pollution test methods are still used today along with monitoring for a number of specific organic compounds (e.g., the TTO list of compounds). However, just like past monitoring programs, testing for the vast majority of chemical pollutants in our water resources is never required.

As a result, Dow was able to discharge the treated effluent to the Tittabawassee River and rely upon a dilution factor of anywhere from 10,000 to 100,000 times to further reduce the concentration of its phenolic discharges. Under these conditions, while phenol would still be a pollutant in the river, the resulting concentration would be so low as to be nearly undetectable. What was more significant, however, was the fact that no other chemicals were specifically treated and removed from this wastewater. Thus, the release of this wastewater would have polluted the Tittabawassee River with an unknown suite of chemicals (see Appendix F) that were not only unregulated but undetected. Given that this approach was typical of how industry[8] dealt with its wastewater, it is not surprising that water resources became grossly polluted.

Individual states continued to regulate municipal and industrial discharges up until the passage of the Clean Water Act in 1972, when national standards for regulating wastewater discharges were incorporated into the National Pollutant Discharge Elimination System (NPDES) permit program[9]. Sadly, this program did not "eliminate" pollution.

Basic Components of the NPDES Program[10]

The Clean Water Act provides the regulatory basis for controlling the quality of wastewater discharges of both toxic and non-toxic pollutants to surface bodies of water through the NPDES permit. This permits covers discharges from municipal and industrial point sources as well as non-point sources of pollution. In setting up the national program, the Clean Water Act required or allowed the following:

- Each state had to establish water quality standards.
- The USEPA had to identify and set discharge controls for toxic compounds (commonly referred to as "Priority Pollutants").
- The USEPA had to define effluent limitation guidelines for municipal and industrial point sources of water pollution.
- Industrial sources of pollution were allowed to be discharged under permit to a municipal POTW.
- All industrial sources under the NPDES permit process had to either adhere to specified effluent discharge limits or pretreat their wastes so as to be in compliance.
- All municipal POTWs were required to have a NPDES permit.
- Each state had to establish programs aimed at controlling non-point pollution (e.g., storm water runoff controls).

These basic features of the NPDES permit program have been in place for approximately 30 years. As of today, all facilities that discharge wastewater

---

[8]This finding is based on the authors' review of the historical records for hundreds of industrial companies in the United States.

[9]Although the NDPES program was a federal regulation, it was implemented at the state level.

[10]NDPES regulations are extremely complex. This discussion is a simplified introduction to the NDPES program and is not intended to be comprehensive.

to waters of a state must have a NPDES permit. Point source permits regulate the amount of pollution that is allowed to be discharged into a receiving body of water. This approach is usually referred to as a "standards" based permit system. In such a system, the discharger can not exceed the set discharge standard for each chemical parameter listed in the NPDES permit. The discharger must also monitor their effluent to demonstrate performance with the permit requirements.

The NPDES program for point sources of pollution as currently implemented allows specific concentrations of regulated chemical pollutants into our water resources as long as these levels are below acceptable criteria. This program also allows for the discharge of an unknown number of unregulated chemicals. Finally, the NPDES program allows permitted facilities to self monitor the quality of their discharges. As a result, the point source NPDES program does not really protect sources of drinking water from pollution. These permits allow a defined level of pollution. In fact, the use of the name "National Pollutant Discharge Elimination System" misrepresents the actual goals of this program. Pollutants are not "eliminated" from our water resources with an NPDES permit, they are only "limited." The real name for this program should be the National Pollutant Discharge Limitation System. Furthermore, this program still relies upon dilution and nature's assimilative capacity to mask the discharge of pollutants.

Point Source Water Pollution and the NPDES Program

The point sources that are regulated by the NPDES program are (1) POTWs that discharge treated wastewater or that use wastewater for spraying or land irrigation and (2) industrial facilities that either discharge to a POTW, discharge directly to a receiving water or use the industrial wastewater for spraying or land irrigation. In each case, the amount of pollution that can be discharged by a given facility is based on the site-specific and region-specific characteristics of the receiving water. For example, the amount of metals that can be discharged depends upon local water quality criteria, the use of the river (e.g., used only for recreation or as a drinking water source), the ability of the river to dilute the metal load and the amount of metals added by other facilities both up and downgradient of the facility in question. Because of these characteristics, it is necessary to discuss the NPDES program with respect to each type of discharge it is designed to regulate.

POTW Discharges: A POTW that receives wastewater from industrial sources has two compliance problems. First of all, the POTW must ensure that the industrial sources meet their pretreatment requirements prior to releasing their wastewater to the POTW. For example, an industrial facility that has an electroplating operation must remove cyanide and metals (such as copper, nickel, chromium and zinc) to levels specified in their permit in order to discharge their wastewater to the POTW. In addition, the electroplating operation must not exceed the Total Toxic Organics (TTO) limit[11]. The TTO is determined by summing of all chemical concentrations greater than 0.01 ppm for the list of compounds presented in Appendix G[12]. A facility may avoid this requirement by certifying that "no dumping of concentrated toxic organic chemicals" into the wastewater will occur [14]. This issue aside, industrial discharges may have effluent limitations for both toxic organic and inorganic compounds to ensure that these discharges do not "upset" the POTW. To assure the POTW of their compliance with these limitations, each industry that discharges to the POTW must monitor their effluents for the chemicals identified in their pretreatment permit and submit these monitoring results to the POTW.

Based on the monitoring data, a POTW can assess civil penalties against industries that do not meet their effluent limitations. Such penalties ensure that industries install and maintain wastewater treatment systems that meet their permit requirements. By using this type of program, the POTWs minimizes the concentration of toxic substances discharged by industrial uses. Some municipalities also set up household hazardous waste collection facilities in order to reduce the amount of hazardous chemicals that households and small businesses dispose via the sewers. Some amount of hazardous inorganic and organic compounds, however, still make their way to POTWs. As a result, POTWs also have to meet effluent limitations. The second compliance problem faced by a POTW is this need to meet their own NPDES permit requirements. Because most primary wastes treated by a POTW are of biological origin, POTW systems are based on the treatment and destruction of bacterial hazards.

At a minimum, most POTWs use basic treatment methods to reduce the concentration of biologic pollutants in their wastewater discharge. These methods removes approximately 85 percent of oxygen-demanding pollutants as measured by BOD from wastewater by using sedimentation and

---

[11]The TTO requirement may be applied to any industry that discharges any of the compounds listed in Appendix G.

[12]Why is this list different from the toxic organic chemicals in primary drinking water standards given in Appendix A? See the "Toxic Dichotomy" discussion in Exhibit 2.2 (*see* pp. 68–69).

Table 2.1
**Chemical Effluent Limitations for the
Central Contra Costa Sanitary District**

Constituent	Daily Maximum	Annual Average
Oil & Grease (ppm)	20	
Ammonia (ppm)	0.16	
Dissolved Sulfide (ppm)	0.1	
Copper (ppb)	19.5	
Lead (ppb)	8.2	
Mercury (ppb)	0.16	
Acrylonitrile	7	
Bis(2-ethylhezyl)Phthalate (ppb)	190	
4,4-DDE (ppb)	0.05	
Dieldrin (ppb)	0.01	
Tributyltin (ppb)	0.06	
Dioxin compounds (mg/year)		9.45

biological treatment (e.g., activated sludge). Depending upon the NPDES permit requirements, a POTW might use (1) advanced treatment methods which remove up to 95 percent of oxygen-demanding pollutants (i.e.,advanced activated sludge methods) and/or (2) tertiary treatment methods to further remove nitrogen and phosphorus compounds from the wastewater. These treatment methods are not designed to remove toxic metals and synthetic organic compounds. Their inability to do so is a significant problem, since not only industrial discharges, but also households and commercial businesses can and do dispose of hazardous materials (i.e., cleaning products, solvents, petroleum products, medications, etc.) into sewers.

Although treatment to remove biological pollutants will result in some portion of toxic metals and synthetic organic compounds partitioning into the sludge1415, biological treatment methods are not meant to remove metals and toxic organics from wastewater. As a result, some portion of soluble metal and synthetic organic chemicals will pass through a POTW and be discharged to a receiving body of water or sprayed on the land. Such discharges can and do pollute groundwater resources [15]. Because of this problem, POTWs also have effluent limitations16. These effluent limitations are usually facility specific. An example of typical effluent limita-

Exhibit 2.2 The Toxic Chemical Dichotomy

The chemicals listed as toxic organic compounds in Appendix G are clearly hazardous to both the operation of a POTW as well as to the aquatic environment that receives polluted wastewater. This fact is obvious since the USEPA actively regulates these compounds. Yet, of the 111 chemicals on the toxic list, only the following chemicals have established primary drinking water criteria:

Benzene, benzo(a)pyrene, carbon tetrachloride, chlordane, chlorobenzene, dioxin, endrin, heptachlor, heptachlor epoxide, hexachlorocyclohexanes (BHC), hexachlorobenzene, 1,2-dichloroethane, dichlorobenzene, 1,1-dichloroethylene, dichloromethane, 1,2-trans dichloroethylene, 1,2-dichloropropane, ethylbenzene, di (2-ethylhexyl)phthalate, pentachlorophenol, polychlorinated biphenyls (PCB), tetrachloroethylene, toluene, 1,2,4-trichlorobenzene, 1,1,1-trichloroethane, trichloroethylene, toxaphene and vinyl chloride.

The remaining compounds on the list do not have established drinking water criteria (i.e., are unregulated). This omission is significant since that TTO compounds, if present in wastewater, will pollute receiving waters, which in turn may be a source of drinking water.

Furthermore, of the 54 organic chemicals listed in the primary drinking water standards in Appendix A, the following chemicals are not listed on the TTO list:

Acrylamide, alachlor, atrazine, carbofuran, 2,4-D, dalpon, DBCP, cis-1,2-dichloroethylene, diaquat, endothall, epichlorohydrin, ethylene dibromide, glyphosate, hexachlorocyclopentadiene, methoxychlor, oxamyl, picloram, simazine, styrene, silvex and xylenes.

tions for a POTW that receives wastewater from both urban and rural areas is illustrated by the NPDES permit (CA0037648) for the Central Contra Costa Sanitary District in Martinez, California.

This treatment plant consists of screening facilities, primary sedimentation, an activated sludge biological treatment process, secondary clarification and ultra-violet disinfection. The plant discharges 45 million gallons of treated effluent per day to Pacheco Slough, which flows to Walnut Creek

(Exhibit 2.2, continued)
These compounds are obviously toxic, yet they would not even be monitored by a POTW*.
The USEPA has promulgated regulations that have defined a group of chemicals as being toxic. One group is considered toxic in drinking water and the other is considered detrimental to POTW operations as well as the aquatic environment. This dichotomy exists because the cost impact of adding more chemicals to the drinking water standard list would be enormous. It would appear obvious to all but the government that if chemicals are defined to be toxic by regulation, they should be added to the drinking water list. To do otherwise confirms the hypocrisy of governmental water quality policies and the urgency of establishing zero pollution goals.
This dichotomy also exists between the United States and the European Union. For example, the European Union proposed in January 2001 to phase out the production of and propose water quality standards for 32 chemical substances**. The following chemicals identified by the European Union are not even listed as a U.S. Primary Drinking Water Standard or on the USEPA TTO list: brominated diphenylether, chlorofenvinphos, chlorpyrifos, diuron, isoproturon, nickel, nonylphenols, octylphenols, pentachlorobenzene, tributyltin compounds and trifluralin. In addition to these chemicals, the Bush administration has endorsed a toxic*** chemicals treaty that will ban the production of 12 compounds in 127 countries. Two of these compounds, mirex and furans, are not listed as a Primary Drinking Water Standard or on the TTO list. Once again, there is no consensus regarding which chemicals should be regulated.

*These compounds occur in urban storm water runoff and from household discharges that are treated by POTWs.
**Water Environment Federation, Vol 13, No. 3 (March 2001).
***White House press release of April 19, 2001.

and then into Suisun Bay. The chemical effluent limitations for this POTW are shown in Table 2.1 (*see* p. 67).

Based on these effluent limitations, the Central Contra Costa Sanitary District Wastewater Treatment Plant must also have a self-monitoring program to determine if they are in compliance with their permit requirements. Under that program, the following chemical parameters must be monitored:

• pH and dissolved sulfides are monitored daily.
• An effluent sample is analyzed monthly for oil and grease, ammonia, cadmium, copper, cyanide, lead, mercury, nickel and tributyltin.
• Quarterly samples are analyzed for arsenic, trivalent and hexavalent chromium, selenium, silver, zinc, 4,4-DDE and dieldrin.
• Samples are analyzed twice a year for dioxin compounds, diazinon and the remaining organic compounds on the TTO list not previously analyzed.

Based on the frequency of this monitoring, there is ample opportunity for toxic pollutants to be discharged from the POTW without detection. In addition to these monitoring requirements, industrial dischargers must also provide monitoring data for nonpriority pollutants (i.e., unregulated toxic pollutants) if the industry "believes"one of these compounds may pass-through the POTW and into the receiving body of water. For example, the Central Contra Costa Sanitary District Wastewater Treatment Plant monitors for diazinon twice a year and tributyltin once a month. However, these two compounds represent only the tip of the "unregulated chemical" iceberg. Although these self-monitoring programs provide some indication of the regulated pollutants that are discharged with the treated wastewater effluent, the pass-through of unregulated organic chemicals in POTW effluents has only begun to be significantly addressed within the last several years.

For example, a study on trace organic compounds in wastewater from Lake Arrowhead, California [16] reported a wide range of both regulated (i.e, Appendix G compounds) and unregulated organic chemicals in their discharge effluent. The compounds found in the Lake Arrowhead study are listed in Appendix H. This list is considerably different than the compounds that are regulated under the NPDES program. Studies conducted at Clemson University [17] analyzed effluents from mixed activated sludge unit for a small suite of organic compounds. Of the chemicals analyzed, those identified in the effluent included acrylonitrile, acrylamide, 4-chlorophenol, dichlormethane, 1,2-dichloropentane, isophorone, methyl ethyl ketone, 2-nitrophenol, 4-nitrophenol, phenol, m-toluate and m-xylene. Another example is a 1995 study by the United States Geological Survey of wastewater pollution in the Mississippi River. This report, previously summarized in Exhibit 1.4, also lists a wide range of pollutants.

Another study in Los Angeles reported that reclaimed wastewater that was injected into groundwater was found to be polluted with EDTA [18]. A United States Geological Survey investigation in Nevada [19] showed the soluble pass-through of pharmaceuticals from septic systems into groundwater. These compounds include chlorpropamide (used for the treatment of diabetes), phensuximide and carbamazepine which are used to treat seizures. A followup to the USGS investigation in Nevada [20] found that the following chemicals passed through the wastewater treatment systems: estradiol, ethynylestradiol, nonylphenol ethoxylates, nonylphenol and octylphenol. It is clear from these studies that a wide range of organic pollutants pass through POTWs and pollute receiving bodies of water. The only questions that remain are (1) how many different man-made organic chemicals are discharged in wastewater effluents and (2) what are their concentrations.

Because the vast majority of these unregulated chemicals are not monitored by POTWs, these chemicals routinely flow undetected into our receiving waters. Furthermore, most POTWs that do not receive industrial discharges still only monitor for indicators of pollution (i.e., BOD and coliform bacteria) and do not routinely monitor for regulated toxic organic chemicals because of economic constraints.

Given the potential for the pass through of chemical pollutants in POTW effluents, injection of these effluents into groundwater aquifers as advocated by various wastewater associations constitutes a serious threat to the public health. This practice is usually termed "Wastewater Reuse." The WateReuse Association contends that [21], "the Colorado River receives the treated effluent from Las Vegas, and the Sacramento/San Joaquin Delta is downstream of the discharge of dozens of Central Valley communities. Several Southern California projects recycle more than 170,000 acre-feet of highly treated effluent every year into underground water supplies used by three million to four million people. Some of these projects have operated safely and reliably for nearly 40 years." In general, wastewater treatment focuses on reducing biological pollutant as opposed to chemical pollution. To say that this practice is safe, when only biologic issues are typically evaluated, is totally misleading.

Scientists have also expressed concern over micropollutants in recycled wastewater. According to researchers at the University of California at Berkeley [8], "the effluent-derived contaminants that are currently being discussed in the scientific community account for only a small fraction of the organic compounds in recycled water. As analytical techniques and our understanding of aquatic and human toxicology improve, it is likely that other chemical contaminants and disinfection byproducts will be detected in recycled water."

Water reuse projects also rely upon the basic chemical nature of unconsolidated sediments in a aquifer, particularly clays and organic matter, to adsorb chemical pollutants that may present in wastewater. Although this process may in fact remove some pollutants from wastewater, these pollutants may be released back into groundwater as a result of natural variations in water chemistry. Furthermore, no studies have evaluated the ability of the sediments in aquifers to attenuate the majority of organic chemicals that can occur in municipal wastewater. As a result, some of these compounds may pass through an aquifer with little or no attenuation, polluting downgradient groundwater extraction wells used for drinking water. A good example of this problem has occurred in Los Angeles, California, where groundwater basins were polluted with NDMA from wastewater recharge [9].

As of January 2002 [22], a total of 26 water reuse facilities using membrane-based technologies were operating in the United States. The degree of pollutant removal achievable using these technologies was not reported so that the overall safety of these treated waters could be evaluated. One thing is clear, however, the use of membrane-based technologies is becoming more and more widespread.

Similar types of pass-through problems occur at industrial facilities that are regulated under permits. In these cases, the problem is much more egregious because many of these industries know the exact chemical specificity of their discharges or have the resources and technical ability to fully characterize their wastewater. They also have the technical knowledge required to design and operate state-of-the-art systems to treat their effluents to virtual purity. Yet, industries continue to discharge regulated organic pollutants at or below discharge standards as well as vast quantities of unregulated organic compounds in their waste streams.

Industrial Discharges: Industrial facilities that have NPDES permits for discharge to either a POTW or directly into the environment must meet at a minimum the published effluent limitations published in the Code of Federal Regulations. While more advanced methods for the treatment of pollutants[15] have been available since the 1950s, the NPDES program still allows the discharge of regulated pollutants to permitted levels, does not monitor unregulated pollutants except in very general consolidated tests and assumes that the environment will assimilate or mask the discharge of both regulated and unregulated chemicals in the receiving body of water.

However, it cannot be assumed that any given environment will assimilate these chemicals. Given the complexity of the organic compounds dis-

---

[15]See Chapter 3 for a discussion of water treatment methods.

Exhibit 2.3 The Fate of Pollutants in Surface Water

When a chemical pollutant is discharged into a surface water, the potential exists for specific chemical, physical and biological changes that can influence the fate of that chemical. In other words, what happens to a chemical once it is released into a river?

The first thing that happens is that the chemical becomes diluted. For example, in some rivers a wastewater that has 500 ppm arsenic can be diluted over 100,000 times to 0.005 ppm. After dilution, the most important factors that influence the concentration of a pollutant in water are those that cause its concentration to decrease. For example, organic compounds that are biodegradable can be converted to carbon dioxide, water and other intermediate chemicals. Furthermore, both organic and inorganic compounds can be removed from surface water by (1) volatilization to the atmosphere, (2) sorption to sediments, and (3) chemical reactions that cause new insoluble compounds to form. Finally, some organic chemicals can be broken down by sunlight.

Ultimately, what must be remembered is that each chemical species, whether organic or inorganic, will most likely be affected by one or more of these factors. Thus, our ability to predict the fate (i.e., concentration in water) of any chemical species requires site-specific environmental and chemical information. Without this information, chemical predictions cannot be reliable.

Chemicals that do not readily biodegrade in the environment (i.e., they tend to persist in the environment) can be a significant hazard if they are toxic and accumulate in aquatic organisms. These environmentally persistent compounds have been identified by the USEPA and are to be either prohibited or significantly reduced in wastewater discharges over the next 10 years. These compounds are alkyl-lead, octachlorostyrene, aldrin/dieldrin, DDT, DDD, DDE, mirex, toxaphene, hexachlorobenzene, chlordane, benzo(a)pyrene, mercury, polychlorinated biphenyls, dioxin and furans. Seven of these compounds, octachlorostyrene, aldrin/dieldrin, DDT, DDD, DDE, mirex and furans, do not currently have drinking water standards.

charged today, scientists and engineers cannot just rely upon dilution and biodegradation to predict the ability of the environment to assimilate a specific compound or compounds. Prediction of the fate of organic and inorganic chemicals in the environment usually requires the use of complex models that are based on either general or site-specific assumptions. An example of the factors that must be considered in predicting the fate of environmental pollutants discharged into surface water resources is given in Exhibit 2.3 (*see* page 73).

Based on pollutant loading data for a watershed and predictions of chemical behavior, an amount of allowable pollution is incorporated into the NPDES permit. For example, the toxic organics that must be removed to "acceptable" levels are usually comprised of the compounds listed in Appendix G and those compounds listed as toxic pollutants in the Code of Federal Regulation, Title 40, Part 129. The Part 129 compounds are aldrin/dieldrin, DDT, DDD, DDE, endrin, toxaphene, benzidine, and poly-chlorinated biphenyls[16]. With effluent limitations established, a given facility must then use an appropriate waste treatment technology to meet its permit conditions on the removal of toxic compounds. Once again, such limitations are facility-specific. An example of effluent limitations for an industrial facility is illustrated using the NPDES permit (CA0005134) for the Chevron refinery in Richmond, California.

This facility discharges an average of 4.0 million gallons per day of wastewater from petroleum refining, petrochemical manufacturing and research, stormwater runoff, construction dewatering at the refinery and pipeline facilities, groundwater monitoring and remediation activities, tank wash water; ship ballast water discharges, and wastewater from storage tanks. The discharge is through a deep water outfall into San Pablo Bay. The chemical effluent limitations for this facility are shown in Table 2.2.

Because of these effluent limitations, the Chevron refinery must also have a self-monitoring program to determine if they are in compliance with their permit requirements. Under that program, the following chemical parameters must be monitored:

- pH is monitored daily.
- An effluent sample is analyzed monthly for oil and grease, total organic carbon, arsenic, total chromium, hexavalent chromium, cadmium, copper, lead, mercury, nickel, selenium, silver, zinc, cyanide and PAH's.
- Quarterly samples are analyzed for ammonia, total phenols and sulfides.
- Yearly samples are analyzed for dioxins and furans, diazinon and the remaining organic compounds on the TTO list not previously analyzed.

---

[16]All of these compounds except DDT, DDD, and DDE are included in Appendix G.

Table 2.2
**Effluent Limitations for an Industrial Facility:
Chevron refinery, Richmond, California**

Constituent	Monthly Average	Daily Maximum
Oil & Grease (lb/day)	1,728	
Ammonia (lb/day)	2,052	
Phenolic Compounds (lb/day)	20.66	
Dissolved Sulfide (lb/day)	30	
Total Chromium (lb/day)	24	
Hexavalent Chromium (lb/day)	1.98	
Lead (ppb)	39.9	
Zinc (ppb)	358.6	
Benzo(a)Pyrene (ppb)	0.94	
Chrysene (ppb)	0.91	
Dibenzo(a,h)Anthracene (ppb)	0.87	
Indeno(1,2,3-cd)Pyrene (ppb)	0.91	
Heptchlor Epoxide (ppb)	0.00132	
Copper (ppb)		14.11
Mercury (ppb)		0.21
Nickel (ppb)		65
Selenium (ppb)		50
Cyanide (ppb)		25
Aldrin (ppb)		0.0014
Alpha-BHC (ppb)		0.13
Benzo(a)Anthracene (ppb)		0.49
Benzo(k)Fluoranthene (ppb)		0.49
Chlordane (ppb)		0.0008
4,4-DDT (ppb)		0.0059
4,4-DDE (ppb)		0.0059
4,4-DDD (ppb)		0.0059
Dieldrin (ppb)		0.0014
Alpha-Endosulfan (ppb)		0.087
Beta-Endosulfan (ppb)		0.087
Endrin (ppb)		0.023
Gamma-BHC (ppb)		0.63
Heptachlor (ppb)		0.0021
Hexachloro-benzene (ppb)		0.0077
PCBs - total (ppb)		0.0007
Toxaphene (ppb)		0.002
Dioxin compounds (ppt)		0.1

Because many industrial facilities discharge a complex mixture of both regulated and unregulated toxic chemical pollutants, the USEPA requires a combination of monitoring methods [23]. These methods, which include chemical specific analyses, bioassessments or whole effluent toxicity tests, are required for most facilities that discharge directly to a water resource. All of these methods have their advantages and disadvantages. For example:

- Chemical Specific Testing is precise but expensive when an effluent contains many toxics. This procedure usually does not analyze for all chemicals and certainly does not consider the interactions between toxic chemicals (as illustrated by the Chevron facility).
- Bioassessments measure ecological effect but do not define the specific chemical source that may be responsible for any damage.
- Whole Effluent Toxicity Testing evaluates the toxicity of the "chemical soup" but offers no information about human health impacts.

Of these methods, only the Whole Effluent Toxicity Testing evaluates the potential synergistic effects of chemical mixtures. This method also evaluates the potential effect of unregulated chemicals independent of whether or not they are identified using chemical specific testing methods. As a result, many industries conduct USEPA-mandated studies to reduce the toxicity of their effluent based on the results of Whole Effluent Toxicity Testing.

The wide application by industry of whole effluent testing is an indirect acknowledgment by the USEPA that a toxic threat does not necessarily occur from one chemical but rather from a mixture of chemicals that may be composed of both regulated and unregulated compounds. This admission is further proof that individual water quality standards are not reliable. Industry use of this method is illustrated in the following examples [24].

A multi-purpose specialty chemical plant operating under a NPDES permit in Virginia was discharging approximately 1.3 million gallons of effluent per day into a surface water resource. This facility manufactured and packaged a number of pesticides. As a result, its effluent contained a mixture of these toxic chemicals. Effluent toxicity testing conducted by the USEPA revealed that the effluent was highly toxic to a number of aquatic organisms. To reduce the toxicity, detailed chemical-specific tests were conducted to identify the compounds responsible for the effluent's toxicity. This testing revealed that the organic chemicals responsible for this toxicity were alkyl diamine, dicyclohexylamine and piperonyl butoxide.

Tosco's Avon Refinery in Martinez, California produces refined petroleum products, primarily gasoline and diesel fuel. Routine NPDES monitoring requirements were limited to pH, ammonia, oil and grease, chromium, zinc, sulfur, chlorine and dissolved oxygen (DO). Because the effluent was found to be toxic, it was tested to identify all organic priority pollutants and major non-priority pollutants. These test data revealed the occurrence of the following compounds in the effluent: ketones, toluene, amines and dibenzofuran.

The Glen Raven Mills in North Carolina used dyes and surfactants. The effluent from its wastewater treatment plant was evaluated using whole effluent toxicity testing methods and found to be toxic to aquatic organisms. With further analysis for specific chemical constituents, it was determined that surfactants, linear alcohol ethoxylate and sodium dodecylbenzenesulfonate were responsible for the toxicity of the effluent. These three examples clearly show that industrial facilities do release unregulated toxic compounds into the environment and that there are no set standards for their control. These tests also show that the toxicity of an industrial effluent is a function of more than one chemical constituent. Both of these facts suggest that water quality cannot be protected by the use of drinking water quality standards as presently promulgated. Furthermore, the use of whole effluent toxicity tests alone are not appropriate as an indicator of drinking water quality since these tests have no direct relationship to human health effects.

More importantly, these examples are significant because the compounds responsible for the effluent toxicity are not listed as toxic organic compounds under the NPDES program (see Appendix G) nor regulated as toxic compounds in drinking water (see Appendix A). This fact illustrates, once again, the failure of existing water quality standards to protect the environment. These sources of pollution are a major source of pollution to our water resources.

However, polluted effluent is easily collected and can be treated prior to its discharge into the environment. Non-point sources of pollution, by their very nature, are dispersed into the environment, are difficult to collect and cannot be easily controlled. As a result, the ability to treat polluted water arising from a non-point sources is very difficult at best and creates a pollution problem that has few practical solutions.

NON-POINT SOURCES OF WATER POLLUTION

Non-point sources of pollution are distributed throughout both urban and rural environment. The major sources of potentially toxic non-point pollution are:

- Pesticide pollution from agriculture land. Pesticides applied to the land can be (1) transported by wind to pollute surface water resources, (2) carried in surface water runoff to adjacent surface water resources or (3) carried in percolating water through soil into groundwater resources.
- Perchlorates from fertilizers.
- Mining of coal and metal sulfides ores (i.e., gold, silver, copper, lead and zinc) from the earth can disturb thousands of acres of land. Because of this disturbance, trace metals and acid will pollute surface and groundwater resources.
- Chemicals released from commercial, industrial and waste disposal facilities pollute surface and groundwater[17].
- Urban environments are subject to the accumulation of chemicals on buildings, streets, impervious surfaces, soils and lawns, in storm channels and sewers, and areas of waste storage or disposal. Common pollutants are pesticides, metals, solvents and petroleum products. During rainfall events, these chemicals will be carried in the runoff to adjacent bodies of water.

Because these sources of pollution tend to provide fairly constant and long-term release of pollutants into the environment, they will always be a threat to water resources and our drinking water quality. As a result, USEPA has supported the development of methods to control non-point source of pollution. However, most of these programs are focused only on controlling pollution arising from sediment and fertilizer releases. Chemical and pharmaceutical pollution remains virtually uncontrolled. The impact of non-point pollution on water quality and our ability to control it is illustrated in the following examples.

Pesticide Pollution

The pollution of water resources by (1) the direct application of pesticides to bodies of water to control aquatic plants, (2) the indirect pollution of surface waters by air transport and/or surface water runoff of pesticides and (3) the pollution of groundwater by pesticide seepage through soil has been

---

[17]Although these sources are usually considered point-sources, many of these sites have multiple points of unreported and unregulated chemical releases that are distributed over a wide area.

well documented since the 1940s. Since then, more pesticides have been developed and spread across both agricultural and urban landscapes. Thus, it is no wonder that much of our water resources are polluted by pesticides or the residues of degraded pesticides.

For example, a recent study conducted by the U. S. Geological Survey and the U.S. Department of Agriculture [25] found that the pesticide diazinon, which is used in agricultural areas of California's central valley, is also present in amphibians that live in pristine mountain ponds and streams. In fact, over 50 percent of the frogs and tadpoles tested at Yosemite National Park had measurable levels of diazinon. As a result, diazinon is suspected as being responsible for the drastic declines in amphibian populations observed in the Sierra Nevada. Since the use of diazinon is restricted to agricultural areas, the diazinion found in the Sierra Nevada ponds, lakes and streams probably is transported on prevailing summer winds from the intensely agricultural San Joaquin Valley.

This type of pollution is virtually impossible to control since the application of pesticides in the form of an aerosol or powder make it susceptible to wind transport. Even after a pesticide has been applied, pesticides adsorbed on fine soil particles can be transported along with the soil eroded by the wind. This type of pollution knows no state or country boundaries. Thus, pesticides banned in the United States but sold to a foreign country may find their way back into the waters of the United States by wind transport.

Pesticides can be applied to a soil or vegetation in the form of a spray, powder, gas or dissolved in irrigation water. Once a pesticide has been applied to a soil it can be transported by surface water runoff[18] to adjacent surface bodies of water. As early as 1956 [26], a study on organic pollution reported that concentrations of DDT in the range of 1 to 5 ppb were found in the drinking water of several cities that used rivers as their sources of supply. This study also concluded that drinking water pollution by organic chemicals was very serious problem since the concentrations at which there might be physiological effects was not then known.

Another early study conducted by the California Department of Water Resources in 1963 characterized the pollution of surface water drainage in the San Joaquin Valley [27]. This study identified the following pesticides in agricultural surface waters: aldrin , BHC, chlordane, CIPC, DDE, DDT, dieldrin, endrin, heptachlor, heptachlor epoxide, lindane, methoxychlor, tedion, thiodan, and toxaphene. Approximately 30 years later, the list of

---

[18]As the result of either rainfall or irrigation.

pesticides that are found in agricultural and urban water resources has increased despite the banning of a number of pesticides. A water quality study conducted by the United States Geological Survey [28] during 1992 to 1996 collected surface water data in 20 of the nation's major hydrologic basins and analyzed for pesticides and pesticide degradation products. The results of this study are summarized in Appendix I. Of the 83 pesticides and pesticide degradation products for which samples were analyzed, 74 were detected. Eleven of the pesticides were detected in more than 10 percent of all surface water samples.

These data demonstrate the widespread distribution of pesticides in our water resources. Given the extensive origins of the pollution, the only effective way to prevent the pollution of drinking water by pesticides is to (1) not allow any runoff into adjacent bodies of water, (2) collect all agricultural runoff for treatment prior to its release, (3) only allow the use of pesticides that biodegrade rapidly so that there is no runoff threat, or (4) ban the use of pesticides completely. Sadly, none of these options is very practical for large areas. Thus, pollution of receiving waters from pesticides in agricultural runoff will continue to be a long-term management problem, particularly since large number of pesticides have been detected in surface waters, yet very few drinking water standards. This finding once again illustrates the inadequacy of drinking water standards, which include only a few pesticides, while ignoring the vast majority.

All pesticides will dissolve in some degree in water and be degraded by microorganisms in the soil root zone. Pesticides can also be adsorbed by soil as the pesticide-laden water percolates downward into the groundwater. Thus, the ability to predict which pesticides are a groundwater pollution threat depends upon the specific properties of each pesticide and the soil environment in which it is used. Nonetheless, it is obvious that a wide range of pesticides currently pollute our groundwater resources and will continue to do so.

Compared to surface water pollution, pesticide pollution of groundwater is an even greater problem because once it occurs it is a long-term hazard, tends to impact large areas, and is expensive to remediate[19]. This impact is typified by the pollution that was created by dibromochloropropane (DBCP) in California. DBCP was added to irrigation water for application to agricultural crops in California's central valley from the late 1950s through the mid-1970s [29]. Studies on the application of DBCP in irrigation water concluded that DBCP moved readily with the

---

[19]Remediation does not necessarily mean the return to the original water quality.

water and was distributed to wherever the water moved [30], including downwards.

As early as 1961 DBCP was reported as being toxic to animals [31]. However, not until DBCP was shown to cause sterility in male workers, did California ban its use in 1977. Two years later, groundwater pollution by DBCP was reported in California's central valley. By 1985, it was determined that almost 2,500 wells throughout the valley had been polluted. It took the California Department of Health Services another four years to establish a maximum contaminant level for DBCP. By 1992, several cities in the central valley had to treat the water at their well heads (i.e., at the point of groundwater extraction) to remove DBCP.

This example illustrates that the public was exposed to DBCP in drinking water resources for decades before any actions were taken. Even though a maximum contaminant level was established, this pollutant is still allowed in drinking water. It is obvious from this example that the use of pollution based standards does not protect drinking water resources.

The extent of groundwater pollution by pesticides is also illustrated in the earlier referenced United States Geological Survey study [28]. The results of this study of ground water pollution is summarized in Appendix J. Of the 83 pesticides and pesticide degradation products analyzed, 59 were detected in groundwater and five were detected in more than 10 percent of all groundwater samples. These studies clearly point out that whether we look at surface waters or ground waters once a resource is polluted there remains a long-term management problem that is not addressed by current drinking water standards.

Perchlorates and Fertilizers

Although perchlorate pollution of surface water and groundwater resources has been primarily associated with rocket fuels [32], the USEPA found in 1999 that garden fertilizers contained perchlorate at concentrations ranging from 1,500 ppm to 8,400 ppm [33]. The source of the perchlorate in fertilizer probably resulted from the use of Chilean nitrates which have been known since the 1800s to contain perchlorate. Because perchlorate is very soluble in water, fertilizers applied to soil can be a source of perchlorate to both surface water and groundwater. The extent of perchlorate pollution from fertilizers is still unknown since widespread monitoring for perchlorate has not yet been required.

Based on the occurrence of perchlorates in Chilean nitrates, it is also possible that perchlorates occur naturally in nitrate deposits in the United

States. For example, a study of natural nitrate pollution in the California's Central Valley [34] identified high nitrate concentrations in the parent materials of the valley's soils. This study suggested that irrigation of these soils contributed to the nitrate groundwater pollution. Given this condition, perchlorates may also have been leached into the groundwater. As a result, perchlorate may also pollute groundwater in regions that contain high nitrate rocks and soils.

Mining and Pollution

When iron sulfide minerals are exposed to the atmosphere and water, the water becomes acid and enriched in iron that eventually leaves yellow and reddish deposits in its wake. Other metals sulfides also react with the atmosphere and water to release toxic metals such as arsenic, cadmium, cobalt, copper, nickel and zinc. While this process has been known since the early 1800s [35], the occurrence of toxic metals, such as zinc, copper, arsenic and antimony in acid mine water was not publicized until the 1870s [36]. By the early 1900s, the hazards of mine waters were well known. For example, a 1907 report [37] on surface water pollution from mining observed that "the appearance of the stream polluted by mine water is striking and somewhat uncanny, for all vegetable and animal life is destroyed, and the bright, clear waters splash forbiddingly over the bed, which is stained yellow by the iron." The types of mines that are responsible for highly acid water and toxic levels of trace metals include all mines that contain metal sulfides (e.g., iron, lead, zinc, copper). Therefore, mines that extract aluminum, coal, copper, gold, lead, silver and zinc will usually pollute surface and groundwater resources with acid and toxic metals. As a result, virtually every state in the nation has some mine-related pollution.

Because the process that creates acid and trace metal pollution is the result of excavation (i.e., bringing metal sulfides to the earth's surface), the minerals that cause the formation of acid can be exposed to the atmosphere for hundreds to thousands of years. As a result, the creeks and rivers that drain mining areas will be polluted far into the future. In addition to polluting surface water, the infiltration of polluted mine waters through fractured rock and soil also impacts groundwater.

The mining process by its vary nature disturbs hundreds of acres of land usually intersected by numerous creeks and rivers. Furthermore, the areas from which acid water is discharged to surface water can occur over numerous locations in a watershed. This fact makes the collection and treatment of the water to reduce acidity and metal concentrations more difficult. In most mining environments, some combination of water management and treatment can be implemented to reduce the pollution. Again it should be stressed that mined lands are only reclaimed, not truly remediated. At best they can be "restored" to some acceptable level of pollution. Thus, even after remedial actions are implemented to reduce pollution, low concentrations of toxic metals can and do occur in downstream water resources.

Industrial, Commercial and Waste Disposal Sites

Virtually every industrial facility in the United States that manufactured or used toxic chemicals has historically polluted air, land and water resources. This pollution is primarily the result of improper waste and chemical handling practices or accidental releases to the environment [38]. Such pollution also occurs at smaller commercial facilities that use or sell chemicals as part of their operations. Good examples of the pollution caused by smaller businesses are dry cleaning facilities that use solvents (e.g., tetrachloroethylene or perchloroethylene) and gasoline stations. Thousands of dry cleaning facilities across the nation have created widespread pollution of groundwater resources with prechloroethylene (PCE) and discharged PCE to sewers where it can then reach surface waters. Similarly, the gasoline additive, methyl tertiary butyl ether (MTBE), is a serious problem in many states. In California, it is a particularly widespread problem. The California Department of Health Services reported in August 2001 that the following sources of drinking water are polluted with MTBE: 0.6 percent of groundwater sources, 4.4 percent of surface water sources and 1.9 percent of public water systems. Although there is no federal water quality standard, the State of California has set a primary drinking water standard for MTBE of 13 parts-per-billion (ppb).

Another problem with MTBE is that it is difficult to remove from water by standard treatment methods. Fortunately, a new treatment method employing both ozone and hydrogen peroxide has been able to remove MTBE below even California's secondary drinking water standard of 5 ppb [39]. Furthermore, the oxidation byproducts of MTBE (e.g., t-butyl alcohol, t-butyl formate, isopropyl alcohol and acetone) are more biodegradable than MTBE itself.

In addition to facilities that either manufacture or use chemicals, waste disposal sites (e.g., sanitary landfills and hazardous waste landfills) that received toxic chemicals have also been significant contributors to the pollution to surface and groundwater. Unlike industrial facilities that manufacture a relatively few number of chemicals, waste disposal sites can contain a vast array of chemicals that can pollute water resources. Thus, the ability to monitor chemical pollution from a landfill is more difficult than monitoring for pollution from industrial facilities because at an industrial facility, regulators generally know what chemical to monitor. As a result, waste disposal sites can also be a source of unregulated chemical pollutants that go undetected by regulatory agencies.

Regardless of the source, when chemicals pollute the environment, the regulations that exist usually require that the extent of the pollution be characterized and remedial measures be implemented. Pollution from industrial, commercial and waste disposal sites around the nation is a significant threat to the environment. It has been estimated that more than 300,000 polluted sites within the United States require some degree of soil and/or groundwater cleanup [40]. The degree to which chemical pollution is remediated is usually based on risk evaluations. Thus, as long as soil pollution is no longer contributing to surface or ground waters, regulatory agencies usually allow the soil to be capped and left in place. Likewise, polluted sediments in streams, rivers and estuaries may not be remediated if (1) the pollution from these sources is not considered a health risk by the regulatory agency and (2) dredging a river may well re-suspend and distribute more pollution downstream. As a result, remediation of pollution at most industrial and commercial sites usually focuses on groundwater.

Groundwater pollution at industrial and commercial facilities is a significant problem because pollution from these facilities tends to be concentrated and there is little, if any, dilution. A plume of polluted groundwater can spread for miles away from the source of the pollution. In many cases it is impossible or impractical, even employing the best available technology, to reduce pollutants to background levels. The most common groundwater pollutants from industrial and commercial facilities tend to be the USEPA's priority pollutants [41].

When groundwater needs to be remediated, most federal and state laws commonly use cleanup goals based on water quality standards. However, the USEPA has recognized that the ability to attain drinking water standard levels using current treatment methods[20] is not feasible at many sites[21] [40]. Furthermore, there is no long-term institutional structure in place to ensure that polluted groundwater will not be unknowingly used.

---

[20]The most widespread method used is called "pump and treat." In other words, pump the polluted water out of the aquifer and treat it before it is discharged back into the environment.

[21]This is especially true for groundwater polluted with chlorinated solvents (i.e., trichloroethylene, perchloroethylene).

Since groundwater pollution persists for a very long time and "pump and treat" systems have been shown to be inadequate at fully addressing aquifer restoration, alternative methods may need to be employed to reduce pollutant levels in the groundwater. Methods commonly employed include:

- Mixing water from unpolluted wells with the polluted well water until the level of the pollutant meets water quality standards.
- Bringing in a new source of drinking water for those areas affected.
- Treating the polluted water at the well head upon its extraction and prior to its distribution to consumers.
- Treating the polluted water at the location of its use (i.e., residents or business).

Regardless of the methods employed, the polluted groundwater remains a non-point source of questionable quality. Even for those few sites where polluted groundwater can be cleaned up to drinking water standards, there will still be a low lingering level of pollution present. As a result, industrial and commercial sites that have polluted groundwater can and do serve as long-term sources of low levels of pollution to receiving waters that ultimately may be used as a source of drinking water.

Urban Runoff

When stormwater scours an urban environment, accumulated debris, residues and wastes are commonly flushed into adjacent bodies of water. Today in urban communities the amount of pollution from industrial facilities and construction sites must be controlled. In larger metropolitan areas (i.e., an urban area with a population above 100,000), runoff is required to be either treated along with domestic wastes at the regional treatment plant or during storm periods, collected and stored in large containment structures, and subsequently treated prior to release. Some communities that do not have the funds necessary to expand wastewater treatment or construct stormwater storage capacity have investigated the process of injecting aluminum sulfate into stormwater conduits to remove pollutants [42]. These studies have shown a 80 to 90 percent removal of heavy metals.

In addition to the more commonly found waste constituents of urban runoff, chemicals that have accumulated in the environment (i.e., roads, roofs of buildings, residential lawns, parks, golf courses, etc.), are carried along in runoff. In some sections of the country, urban runoff is viewed as a "water resource" of value and is therefore directed to percolation basins which allow it to percolate through the soil and into groundwater, thereby recharging the groundwater aquifers. For this process to be effective, the soil must retain the pollutants so that ground water resources are not degraded.

However, recharging aquifers with polluted surface water simply shifts the pollution from surface water to the groundwater. For example, residential drinking water wells have been polluted by pesticides commonly used in urban environments [43]. This type of pollution typifies the hazard of leachable chemicals that are commonly found in the urban landscape.

BASIN WATER QUALITY MANAGEMENT

In an attempt to improve national water quality, the Clean Water Act (CWA) established regional performance-based programs to manage the water quality of regional watersheds (e.g., NPDES program, storm water management programs and the control of non-point pollution). As part of these programs, the CWA also established the Total Maximum Daily Load Program (TMDL). Under this program, a given watershed may need to place a TMDL on specific organic pesticides in a watershed that does not meet the established water quality criteria. In general, this program requires states to (1) identify those bodies of water that do not meet applicable water quality standards and (2) provide a road map for efforts to attain and maintain state water quality standards. This usually means that pollution controls need to be improved to meet existing water quality standards.

The ability to identify the pollution parameters that need to be controlled is usually not a problem. The current causes of chemical water quality impairment include nutrients, metals, dissolved oxygen, toxic organics, mercury and pH. Almost all of these pollutants do come from non-point pollution sources [44]. The real problem associated with this program is the ability to quantitatively determine to what extent each pollution source needs to be controlled (i.e., what is the maximum daily pollutant load allowable from each source), the technical ability to reduce the contributions from non-point sources, and the ability to impose potentially economically crippling costs on both agricultural and municipal communities in order to comply with the CWA.

Although it is important to control the amount of pollutants discharged into the environment, the TMDL program still relies on performance-based pollution standards. In other words, it is acceptable practice to pollute the environment up to specified water quality standards and not worry about unregulated chemicals. As a result, we will always have polluted water resources that we may ultimately want to use as sources of drinking water.

POLLUTION SOURCES AND WATER QUALITY

In today's chemically dependent society, pollutants discharged to the environment can not be assumed to be assimilated to a degree where there will be no impact on water quality. Since the turn of the 20th century, pollution has become increasing complex with a diversity of chemicals that have been shown to persist in the environment. When this fact is combined with a multiplicity of pollution sources found in both urban and rural landscapes, the potential for the pollution of drinking water resources will continue to increase. The boundaries between wastewater and drinking water have blurred to the point where most Americans are impacted by the fact that their drinking water supplies originating from sources polluted by wastewater [45].

Our drinking water resources are threatened by a vast array of chemicals. This threat exists because of the following set of conditions:

- Chemicals released into the environment are not necessarily removed from water by natural processes. Yet, scientists and engineers depend upon these natural processes to protect drinking water without specific proof that surface and groundwater systems can safeguard our water resources from the vast array of organic chemicals that are introduced into the environment.
- Point sources that discharge wastewater under the NPDES permit system are allowed to discharge regulated pollutants at specified levels. Thus, low concentrations of pollution are legally discharged to receiving waters.
- Point sources that discharge wastewater under the NPDES permit system are also allowed to discharge unregulated pollutants to receiving waters.
- The removal of unregulated pollutants from POTW discharges can most likely only be accomplished by using microfiltration and reverse osmosis technologies [8]. Use of these technologies is more frequent in water restricted states such as California and Arizona [46] but these technologies have had limited use in the rest of the nation because of their cost.
- Non-point sources are a continuing source of pollution to receiving waters. These sources of chemical pollution will continue to be a significant problem for water resource managers and regulatory agencies.

- The list of pollutants that are not regulated in drinking water and have already been identified in our water resources is extensive. This list, which is presented in Appendix K, includes a large number of compounds. However, these compounds only represent a fraction of the compounds that are probably in our water resources.
- No consistent list of pollutants, which should be regulated, has ever been compiled. For example, the State of California under Proposition 65 was required to publish a list of chemicals known to cause cancer or reproductive toxicity. This list is given in Appendix L. When the chemicals listed in Appendix J are compared to toxic compounds regulated in wastewater (Appendix G), the toxic pesticides regulated in food (Appendix E) and the Primary Drinking Water Standards (Appendix A), it is absolutely clear that the current list of drinking water standards is meaningless.

Because of these conditions, the chemical purity of our drinking water sources cannot be guaranteed and is not safe. Unfortunately, the extent and magnitude of the problem will never be known since only a handful of unregulated pollutants in drinking water are actually monitored.

Therefore, depending upon where one lives, they should be aware that the source of their drinking water may be significantly polluted. For example, people who obtain drinking water either downstream or downgradient from the following activities should be concerned about consuming chemical pollutants (in a decreasing order of severity):

- Surface waters that transect crop-producing areas that use pesticides.
- Surface waters that receive treated industrial discharges or mine runoff.
- Surface waters that receive discharges from a POTW.
- Surface waters that receive urban runoff.
- Groundwater that receives effluent from sanitary leach fields.
- Groundwater that is recharged with treated municipal or industrial wastewater effluent.
- Groundwater that is under the influence of surface waters that receive agricultural runoff, industrial or POTW effluents.

Unfortunately, these conditions existing over most of the United States. Communities that receive their drinking water primarily from natural mountain rainfall and snowmelt have the potential of obtaining the least polluted water. However, it should be remembered that even mountain lakes and rivers are polluted by pesticides transported hundreds of miles in air.

References

1. E. C. Clark, "Report on the Main Drainage Works of the City of Boston," *Annual Reports of the Board of Health* (1885).

2. Mason, *Water-Supply*, John Wiley & Sons (1896).

3. Goodell, Edwin B., "A Review of the Laws Forbidding Pollution of Inland Waters in the United States," U. S. Geological Survey, Water-Supply and Irrigation Paper No. 103 (1904).

4. Parker, Horatio N., "Stream Pollution, Occurrence of Typhoid Fever, and Character of Surface Waters in Potomac Basin," U. S. Geological Survey Water-Supply and Irrigation Paper No. 192 (1907).

5. Bruce, Mark, et al., "Mercury," *Water Environment and Technology*, Vol. 13, No. 11. (November 2001).

6. Anonymous, "WE&T Waterline, Tracking Estrogen Pollution," *Water Environment and Technology*, Vol. 13, No. 11. (November 2001).

7. Daughton, Christian and Thomas Ternes, "Pharmaceuticals and Personal Care Products in the Environment: Agents of Subtle Change?" *Environmental Health Perspectives*, Vol. 107, Supplement 6 (December 1999).

8. Sedlak, David, James Gray and Karen Pinkston, "Understanding Microcontaminants in Recycled Water," *Environmental Science and Technology*, Vol. 34 (December 2000).

9. California Department of Health Services, California Drinking Water: NDMA-Related Activities, updated December 6, 2001.

10. Harlow, I. F., T. J. Powers and R. B. Ehlers, "The Phenolic Waste Treatment Plant of the Dow Chemical Company," *Sewage Works Journal* (November 1938).

11. Powers, Thomas, "The Treatment of some Chemical Industry Wastes," *Sewage Works Journal* (March 1945).

12. Dow Chemical Company, *Dow Industrial Chemicals and Dyes*, Midland, Michigan (1938).

13. United States Public Health Service, "Public Health Service Drinking Water Standards," *Public Health Reports*, Vol. 61, No. 11 (March 15, 1946).

14. Code of Federal Regulations, Title 40, Part 413, Electroplating Point Source Category (7-1-00 edition).

15. Bouwer, H., P. Fox and P. Westerhoff, "Irrigating with Treated

Effluent," *Water Environment and Technology* (September, 1998).

16. Levine, B. B., et al., "Treatment of Trace Organic Compounds by Ozone-Biological Activated Carbon for Wastewater Reuse: The Lake Arrowhead Pilot Plant," *Water Environmental Research*, Vol. 72, No. 4 (July/August 2000).

17. Magbanua, Benjamin S., et al., "Quantification of Synthetic Organic Chemicals in Biological Treatment Process Effluent Using Solid-Phase Microextraction and Gas Chromatography," *Water Environmental Research*, Vol. 72, No. 1 (January/February 2000).

18. Barber, L. B., et al., "Organic Constituents that Persist During Aquifer Storage and Recovery and Reclaimed Water in Los Angeles County, California." In *Conjunctive Use of Water Resources-Aquifer Storage and Recovery*, ed. D. R. Kendall, 261–272, Herndon, Virginia: American Water Resources Association (1997).

19. Seiler, Ralph, Steven Zaugg, James Thomas and Darcy Howcroft, "Caffeine and Pharmaceuticals as indicators of Wastewater contamination in Wells," *Groundwater*, Vol. 37, No. 3 (May/June 1999).

20. Roefer, Peggy, et. al., "Endocrine-disrupting chemicals in a source water," *Journal of the American Water Works Association*, Vol. 92, No. 8 (August 2000).

21. Saunders, Robin G., "Letters to the Editor; What You Don't Think You Wanted to Know About Recycled Wastewater," *Scientific American*, Vol. 284, No. 6 (June 2001).

22. Crook, James and William Vernon, "A Clear Advantage, Membrane Filtration is Gaining Acceptance in the Water Quality Field," *Water Environment and Technology* (January 2002).

23. U.S. Environmental Protection Agency, "Introduction To Water Quality-Based Toxics Control for the NPDES Program," Office of Water, Washington , D.C., EPA 831-S-92-002 (March, 1992).

24. U.S. Environmental Protection Agency, "Generalized Methodology for conducting Industrial toxicity Reduction Evaluations (TREs)," Risk Reduction Engineering Laboratory, Cincinnati, Ohio, EPA 600-2-88-070 (April 1989).

25. U.S. Geological Survey, "USGS Research Finds that Contaminants May Play and Important role in California Amphibian Declines," News Release of 12-7-00 from USGS, MS119 National Center, Reston, VA 20192.

26. Middleton, F. and A. Rosen, "Organic Contaminants Affecting the Quality of Water," *Public Health Reports*, Vol. 71, No. 11(November 1956).

27. California Department of Water Resources, "San Joaquin Valley Drainage Investigation," *Bulletin* No. 127 (April 1969).

28. U.S. Geological Survey, "Pesticides in Surface and Ground Water of the United States: Summary of Results of the National Water Quality Assessment Program" (July 22, 1998).

29. University of California at Davis, "Groundwater Quality and Its Contamination from Non-point Sources in California," Centers for Water and Wildland Resources, Water resources Center Report No. 83 (June 1994).

30. O'Bannon, J., "Application of Emulsifiable Digromochloropropane in Irrigation Water as a Freeplanting Soil Treatment," *Plant Disease Reporter*, Vol. 42, No. 7 (July 15, 1958).

31. Torkelson, T, Sader S. and C. Rowe, "Toxiclogic Investigations of 1,2-Dibromo-3-Chloropropane," *Toxicology and Applied Pharmacology*, Vol. 3 (1961).

32. USEPA, "Region 9 Perchlorate Update," USEPA Region 9, San Francisco, California (June 1999).

33. Renner, Rebecca, "Study Finding Perchlorate in Fertilizer Rattles Industry," *Environmental News, Environmental Science and Technology*, Vol. 33, No. 19 (October 1999).

34. Sullivan, Patrick J and Jennifer L. Yelton, "An Evaluation of Trace Element Release Associated with Acid Mine Drainage," *Environmental Geology and Water Science*, Vol. 12 (1988).

35. Grammar, John, "Account of the Coal Mines in the Vicinity of Richmond, Virginia," *The American Journal of Science*, Vol. I. (1819)

36. Williams, Chas. P., "Analysis of Mine Water from the Lead Region of South-west Missouri," *American Chemist*, January (1877).

37. Parker, Horatio N., "Stream Pollution, Occurrence of Typhoid Fever, and Character of Surface Waters in Potomac Basin," U.S. Geological Survey, Water-Supply and Irrigation Paper No. 192. (1907).

38. Liang, Sun, et. al., "Treatability of MTBE-contaminated groundwater by ozone and peroxone," *Journal of the American Water Works Association* (June 2001).

39. Sullivan, P., F. Agardy and R. Traub, *Practical Environmental Forensics, Process and Case Histories*, John Wiley & Sons (2001).

40. National Research Council, *Alternatives for Ground Water Cleanup*, National Academy Press, Washington D.C. (1994).

41. Montgomery, J. and L. Welkom, *Groundwater Chemicals Desk Reference*, Lewis Publishers, Chelsea, Michigan (1990).

42. Herr, J. L. and H. H. Harper, "Reducing Nonpoint Source Pollutant

Loads to Tampa Bay Using Chemical Treatment," *Water Environment Federation* (October 2000).

43. Eitzer, B. D. and A. Chevalier, "Landscape Care Pesticide Residues in Residential Drinking Water Wells," *Bulletin of Environmental Contaminant Toxicology*, Vol. 62 (1999).

44. USEPA, "The National costs of the total Maximum Daily Load Program (Draft Report)," Office of Water, Washington, DC, EPA 841-D-01-003 (August 2001).

45. Maxwell, Steve, "Ten Key Trends and Developments in the Water Industry," *Journal of the American Water Works Association*, Vol. 93, No. 4 (April, 2001).

46. Freeman, Scott, G. F. Leitner, J. Crook and W. Vernon, "A Clear Advantage, Membrane filtration is gaining acceptance in the water quality field," *Water Environment and Technology* (January 2002).

Chapter 3

Precaution

"To be wholesome, water must be . . . free from poisonous substances. The possibility of their presence in water supplies is sometimes unsuspected."

— Gordon Fair and John Geyer,
Water Supply and Wastewater Disposal, 1954

Water for domestic consumption has been collected, stored, treated and distributed to thirsty consumers as far back as recorded history. Today nothing has changed except that the consumer has a greater selection of water products from which to choose. Drinking water is now available from domestic water resources taken from surface and underground supplies, and bottled water from both domestic and world-wide distributors. Given the number of water products available, one would assume that an informed consumer would be able to get water that is sparkling clear, good tasting and free from "poisonous substances." Unfortunately, the consumer has no real assurance that the water they drink does not contain known or "unsuspected" pollutants. In fact, the reverse is true. Drinking water can be assumed to be polluted unless proven otherwise.

Because drinking water contains chemical pollutants that have no established direct relationship to human health, no reliable means exists for establishing their risk to consumers. Additionally, current USEPA programs to identify and control chemical pollutants in drinking water are ineffective. Thus, with "unsuspected" pollutants in our drinking water and in the absence of credible scientific evidence to support the use of health based water quality standards for known toxic water pollutants, consumers should take to heart the goal of the "Precautionary Principle." This principle maintains that in the absence of scientific evidence of harm, measures should be taken to protect the public health. The same principle is also applicable to the intentional poisoning of drinking water resources as the result of a terrorist attack.

For consumers who wish to exercise precautionary measures, there is only one real option. Do not allow yourself to be exposed to chemical pollutants in water. However, this is not an easy task given the multiple routes of chemical exposure we all face each and every day. Because pollutants enter the body by breathing volatilized chemicals while showering, are adsorbed through the skin while bathing and are ingested, a consumer cannot reduce their exposure just by placing a water filter on the tap of the kitchen sink or by drinking bottled water. The solution requires both individual actions to reduce exposure in the home and workplace as well as supporting governmental policies to reduce chemical pollutants in drinking water and bottled beverages.

The only sure way to minimize the threat to human health from chemical pollution in our drinking water is to adopt a program that will implement the use of the best available treatment technologies by community water systems and in-home and workplace water treatment systems for those who have their own private water source.

BASIC WATER TREATMENT

There is clear evidence that ancient societies were concerned with water quality as far back as 2000 BC. These societies improved water quality by allowing the natural action of soil, sand and course gravel to filter water, by boiling water, and by allowing water to settle in reservoirs and basins in order to remove suspended solids. Today at modern water treatment plants, essentially the same technology is employed, although it is refined and much better understood. Whether we look back 4000 years or look forward into the future, certain truths appear to be consistent about water quality:

- If water has a bad taste or odor, people will not drink it.
- If water shows evidence of turbidity (i.e, it's muddy) and color, people will not drink it.
- If water makes people sick they will stop drinking it.

Thus, water treatment usually strives to produce water that does not have bad taste, odor, color or turbidity, and is free from bacterial threats.

The basic elements of a typical water treatment plant include most if not all of the following processes in order to meet the previously mentioned objectives: pretreatment, prefiltration, filtration, chemical treatment and disinfection.

- Pretreatment of raw water usually consists of (1) removing floating debris such as weeds and leaves by using screens, (2) aeration such as one often sees in fountains to remove many of the more volatile chemicals that can add taste and odor to water and to add oxygen to the water (i.e., low oxygen in water often leaves a "flat taste"), (3) sedimentation to remove dirt and other materials heavier than water and (4) occasionally, chlorination to both partially disinfect the water as well as oxidize and remove some organic chemicals from the water.
- Prefiltration usually consists of adding chemicals to the water to flocculate[1] and filter out suspended particles. However, the added chemicals (acrylamide and epichlorohydrin) can also be carried along with the treated water when it is distributed to the consumer.
- Filtration is generally accomplished by the use of sand beds. Water passing through sand will remove the vast majority of the suspended and colloidal materials found in water.
- The final water treatment step usually consists of chlorinating the water to eliminate bacterial pollution and to prevent the regrowth of bacteria in the distribution systems carrying the treated water to the consumer. In addition to chlorination, the water may also be treated to reduce hardness (i.e., remove excessive amounts of calcium and magnesium) or the pH may be adjusted to prevent corrosion or scaling of the water distribution mains.

These treatment methods, however, were not specifically designed to remove toxic metals and organic compounds. It is only by chance that some toxic metal and organic compounds are removed from the raw water by the oxidation, coagulation and filtration processes. Thus, toxic metal and organic compounds can and do pass through the water treatment plant and into the distribution system to the consumer.

Although treatment is necessary to improve the quality of drinking water, the process also creates its own pollution problems. The standard methods of water treatment employed by water utilities can add suspected carcinogens to treated water by (1) the addition of flocculents to remove suspended solids and (2) by the use of chlorine, chloramine or bromine to destroy biological hazards and the subsequent creation of disinfection byproducts.

---

[1] A process that causes very fine particles to combine into larger particles.

Flocculents

In order to remove small suspended particles in water, a flocculent is used to coalesce small particles into a larger particle so that these particles can be picked up by a filter. The two chemical polymers used for this purpose are acrylamide and epichlorohydrin. Both of these compounds are listed as "reasonably anticipated to be human carcinogens" in the USPHS Ninth Report on Carcinogens. These compounds also have established limits in drinking water (see Appendix A). In other words, drinking water may routinely contain low levels of these anticipated carcinogens.

Disinfection By-Products

Chlorine, chloramine and bromine are used in water treatment plants as a disinfectant to destroy biological hazards. Unfortunately, chlorine and bromine can undergo complex chemical reactions with either natural or synthetic organic chemicals and create newly chlorinated or brominated organic compounds in the finished water. The compounds of concern are halogenated methanes, haloacetic acids and nitrosamines.

When natural waters are either chlorinated or brominated, over 100 potentially toxic halogenated compounds can be created [1]. Of these potentially toxic compounds, only chloroform, bromoform, bromodichloromethane and dibromochloromethane have established Primary Drinking Water Standards. These compounds as a group are known as the total trihalomethane (TTHM) compounds and should not exceed 100 ppb in drinking water. Of the TTHM compounds, only chloroform and bromodichloromethane are listed as "reasonably anticipated to be human carcinogens" in the USPHS Ninth Report on Carcinogens.

However, the Stage 2 Disinfection By-Product Rule issued in May 2002 will (1) lower the TTHM Maximum Contaminate Level (MCL) to 80 ppb, (2) establish a MCL of 60 ppb for five Haloacetic acids (HAA5), including monochloroacetic acid, dichloroacetic acid, trichloroacetic acid, monobromacetic acid and dibromoacetic acid, (3) establish a MCL of 10 ppb for bromate, and (4) establish a MCL of 1000 ppb for chlorite. Even with these modifications, a recent article in the Journal of the American Water Works Association [2] reports that there should be standards for the nine haloacetic acids (HAA9) instead of the HAA5 compounds. This article states that "given the HAAs are thought to pose greater health risks than [T]THMs do, the possible widespread occurrence of HAA9 should be of concern to water suppliers and regulators alike." This finding should also be of concern to the average consumer.

According to the report entitled, "Consider the Source, Farm Runoff, Chlorination Byproducts, and Human Health" by the Environmental Working Group, halogenated methanes and haloacetic acids represent a serious health threat to the American Public and specifically pregnant women [1]. Some of the report's main conclusions are that more than 137,000 pregnancies each year are at increased risk of miscarriage and birth defects and of the 50 communities studied the top five with the most risk were Montgomery County, Maryland; Washington, DC; Philadelphia; the suburbs of Pittsburgh; and San Francisco.

Furthermore, the Environmental Working Group report contends that the USEPA's ability to link one health effect (i.e., bladder cancer) to halogenated byproducts illustrates how the nation's health tracking programs force decisions based on just a fraction of the public health data on environmental pollutants. To better quantify the health impacts of halogenated byproducts, they recommend the creation of a national health tracking system. In other words, not only does the USEPA lack the ability to establish human health effects for specific chemicals, it also lacks the ability to track environmental impacts on individuals and communities.

Even though the Stage 2 Rule will establish new drinking water standards for the halogenated disinfection byproducts, the same flaws will remain. Low levels of a handful of suspected halogenated compounds will be allowed in drinking water, while virtually hundreds of potentially toxic halogenated compounds go unregulated. Moreover, in order to meet just the new HAA5 standards, new treatment technologies will most likely have to be installed by many community water systems.

One set of disinfection byproducts not addressed by the Stage 2 Rule are nitrosamines. Research suggests that nitrosamines are formed as the result of the disinfection process [3]. Nitrosamines have also been identified in drinking water distribution systems as a result of chlorination. Nitrosamine formation within the drinking water system is a significant threat because these compounds are potential carcinogens. One nitrosamine, N-nitrosodimethylamine (NDMA) is of particular concern because of its carcinogenicity and widespread distribution [4].

In addition to the occurrence of water treatment chemicals and disinfection by-products in community water systems, fluoride is often added to the finished water product. The debate over the benefits of the fluoridation of drinking water to reduce dental cavities versus its carcinogenic and toxic characteristics has gone on for several decades. It is not our intent to continue this debate other than to point out that fluoride is a toxic substance. According to the 1960 Merck Index [5], sodium fluoride[2] is an insecticide used to kill roaches and ants. Thus, in the realm of polluted drinking water, fluoride is just another toxic compound among many that can be found in drinking water. Moreover, fluoridation to a concentration not to exceed 4.0 ppm of fluoride is consistent with governmental policies, which allow drinking water to be polluted with toxic chemicals based on the reliance on "standards" to maintain safety and protect human health.

Man's use of water treatment technologies to purify water has proven to be a significant factor in protecting the public health. For waters that come from a protected watersheds[3], chemical pollution is far less of a concern than water from urban or agricultural watersheds. In addition to the chemical pollutants already present in the water resource, basic water treatment technologies can add a number of potentially toxic man-made chemicals to a community water supply. These concerns were raised in a *Consumer Reports* [6] article in 1974 entitled, "Is the Water Safe to Drink?" This report concluded that (1) no substance capable of causing cancer in any animal species at any dosage should remain in the water we drink if a feasible method is available for removing it and (2) the cost of promptly improving our drinking water is reasonable enough to justify the added protection that would be gained. Sadly, almost 28 years later these recommendations have yet to be implemented even though various treatment technologies are capable of removing pollutants to as close to zero as possible.

BEYOND BASIC WATER TREATMENT

The ability for a water treatment facility to deliver a wholesome and high quality product to its consumers usually occurs when advanced treatment methods are used. Some of the more advanced technologies employed today focus on the ability to use instrumentation to control the optimal quantities of chemicals needed in process operations (i.e., coagulation, flocculation, sedimentation and filtration). Such instrumentation is also used to continuously monitor water quality throughout the water treatment process. In addition to instrumental controls, numerous advanced treatment methods can be employed to reduce both biologic and chemical pollutants.

The use of advanced instrumentation in water treatment systems is an obvious benefit to a water utility because these technologies save money on chemicals and labor. As a result, these advance instrumentation meth-

---

[2]The specific chemical compound used to fluoridate water.

[3]Such watersheds have little or no human activity so that chemical and human pollution is minimized although not absent.

ods have been installed at the water treatment plant. The use of advanced treatment methods does, however, increase the operating cost of water treatment systems. Because cost is the limiting factor for the use of advanced treatment methods, their use tends to be restricted to those few facilities that can afford the added expense. For the facilities that can pay the added costs, the following advanced treatment methods are available:

- In the United States, activated carbon or granulated activated carbon (GAC) has been used to remove organic chemicals from water since the early 1900s.
- Ion exchange resins have been employed to remove inorganic metals for decades.
- Ozonation has been successfully used in Europe instead of chlorine and bromine to both disinfect water and oxidize organic constituents. Such systems are still being tested in the United States and as yet have not received wide acceptance. Ozonation in combination with hydrogen peroxide (i.e., advanced oxidation) has been used in the United States to remove persistent compounds such as MTBE [7].
- Ultraviolet light, which has also been employed in Europe, is presently being evaluated at a number of municipal treatment facilities in the United States for bacterial and viral control.
- As far back as 1985, the French have made use of membrane separation technology (i.e., reverse osmosis) for the removal of organic and inorganic compounds. Today a number of reverse osmosis installations are operating in the United States. Reverse osmosis membranes can hold back a wide range of micro pollutants including pesticides and pathogenic organisms and can replace the need for both ozonation and activated carbon filtration. A variation of reverse osmosis is called nanofiltration, which operates at lower pressures than reverse osmosis, can remove compounds in the 300 to 1,000 molecular weight range.
- Advances in the development of organic polymers for improved flocculation are referred to as enhanced coagulation.

However, due to the age of most community water supply systems (50 to 100 years old), local state and federal funds tend to go towards existing maintenance problems[4] instead of advanced treatment technologies. For

[4]Older systems suffer from clogged reservoirs, which reduces their storage capacity, leaking reservoirs, which lead to extensive water losses, and deteriorating treatment facilities and transmission pipes.

example, the City and County of San Francisco was recently faced with a major upgrade of its transmission pipelines bringing water from the Hetch Hetchy reservoir to local storage reservoirs to meet seismic codes. The estimated upgrade and retrofit costs will run into the billions and will necessitate rate increases for many years. Yet, this upgrade only deals with the main transmission system.

Although advanced technology does exist, it is infrequently used on most of the public water systems and rarely on the smaller rural systems. Yet, a state-of-the-art treatment plant that used all available technologies to remove disease causing constituents such as bacteria and viruses, removed all naturally occurring and synthetic organic chemicals, and limited the amounts of naturally occurring metals and radioactive elements could effectively protect human health as well as maintain the integrity of the distribution system.

The state-of-the-practice is to "patch up" what is broken or in need of repair, update existing treatment units and, when forced by either costumer complaint or regulatory requirement, add additional technology in hopes of maintaining or keeping up with regulatory compliance. The magnitude of the facility maintenance problem was recently described in an USEPA study [8]. According to the USEPA, an estimated expenditure of over $1 trillion will be needed between 2010 and 2020 to meet current objectives of upgrading and extending the operational capabilities of our nation's water and wastewater systems. Quoting the USEPA, "currently capital spending is not adequate, new investment is flat, the long term financial solvency of many systems is doubtful and more households are having problems affording services."

Government funding and regulatory requirements aside, a community water system could raise the cost of its product to its consumers. All that is needed is political approval from the appropriate regulatory agency (e.g., mayor, city supervisors, public utility board). When there is a clear threat, community water systems do these advanced water treatment technologies. The problem is that there is no established process for determining that a threat of chemical pollution justifies the cost of implementing advanced treatment technologies. Furthermore, even if a threat was identified, not all community water systems would be politically and financially able to respond. For example, privately owned water utilities might be more likely to implement advanced treatment technologies simply because they operate only one business. This situation is in sharp contrast to municipally owned water utility where the owner (a city or county) have competing budgets and funding priorities.

In the western United States, the cost of water treatment for both biological and chemical components is approximately 15 percent of a community water system budget. To install the best available technologies to remove chemical pollutants to a level as close to zero as possible would only increase these budgets by 15 to 25 percent [9].

Given the actual cost to the consumer and the quality of water received, many communities would probably be agreeable to such a increase in the cost of water. This conclusion assumes that either (1) some level of governmental support will be required to subsidize low income water users and/or (2) the increased costs are weighted towards to those consumers that have the greatest consumption. If community water systems are not willing to implement all the technologies required to provide a water product with an absolute minimum concentration of chemical pollutants, then the necessary technology will have to be phased in over a number of years. For example, compliance with the Stage 2 Disinfection Byproduct Rule will take approximately six to twelve years. Under this program, the community water systems that cannot meet the TTHM and HAA5 Maximum Contaminant Levels will have to use the best available technology, such as enhanced coagulation, GCA or nanofiltration, if the water treatment system uses chlorine as the primary and residual disinfectant.

Under all of these conditions, it is clear that many community water systems will have to add advanced treatment technologies in order to comply with the Stage 2 Rule. However, it is not clear how many community water systems would be willing to upgrade their treatment system to the point that their product has the minimum concentration of chemical pollutants possible without governmental funding. Regardless of funding issues, the economies of scale would suggest that the larger municipalities would have a greater probability of installing advanced treatment methods for their consumers than small systems (e.g., those serving less than 100,000 persons).

AN ISSUE OF EQUALITY

The size and type of a water treatment system determines two important characteristics that can influence the quality of the water a customer consumes. These characteristics are the extent to which advanced treatment technologies were used and the type and frequency of water quality monitoring conducted.

Because the implementation of advanced treatment technologies is a function of local, state and federal funding, many water systems can only provide basic water treatment. In general, the larger the water provider, the greater probability that advanced treatment technologies may be used. Thus, consumers who receive their water from a provider that serves a population over 100,000 (i.e., large system) have the greatest probability of their water being treated to the highest quality. As of 1993 [10], approximately 45 percent of the population of the United States was served by large water systems while roughly 2.5 percent were served by small water systems (i.e., those that serve less than 500 people). These figures, however, do not include the approximately 22 million individuals who obtain their water from private water resources. Thus, a significant proportion of the population receives a water product that could and should be treated to a greater degree.

The other significant factor that can influence whether or not a customer will knowingly or unknowingly consume polluted drinking water is the extent and frequency of water quality monitoring that is required. Federally mandated monitoring requirements are tied to a water provider's classification and size. Therefore, it is necessary to define the USEPA's public water system classification scheme before monitoring requirements can be defined. The USEPA has defined a public water system as a facility that provides piped water for human consumption to either a minimum of 15 service connections or serves at least 25 persons for a minimum of 60 days a year. Based on this definition, the USEPA has further defined the following three types of public water systems :

- Community Water System (CWS): A public system that supplies water to the same population year-round (e.g., individual residences or businesses).
- Non-Community Water System (NCWS): A public water system that regularly supplies water to at least 25 of the same people at least six months per year, but not year-round (e.g., schools or hospitals).
- Transient Non-Community Water System (TNCWS): A public water system that provides water in a place where people do not remain for long periods of time (i.e., a campground or highway rest area).

Because CWS systems generally provide a water product to locations where individuals permanently reside, the quality of the water is measured by drinking water standards that are supposed to protect against potential heath effects that may occur from long-term exposure[5]. Thus, in order for

---

[5]The NTNCWS and TNCWS systems mostly serve individuals who have another primary source of drinking water. As a result, the USEPA generally allows these systems to only monitor for pollutants that may have an acute or immediate health effect on the consumer (e.g., microbiological hazards).

water quality standards to serve as the means for protecting human health, the chemicals with drinking water standards must be monitored in the water provided to a community.

CHEMICAL MONITORING AND
WARNINGS FOR REGULATED POLLUTANTS

The Federal Safe Drinking Water Act that was passed into law in 1974 attempts to ensure the quality of drinking water by setting both standards (Appendix A) and the frequency of sampling drinking water for analysis. In other words, by setting water quality standards, monitoring will be conducted to determine if a water resource does not exceed any standard and, if the standard is exceeded, these data can be used to protect the public health. The monitoring requirements developed by the USEPA to protect human health can be reviewed in Appendix M. Unfortunately, these monitoring regulations do not really protect the public health because of the following defects:

• Only chemicals that are listed in Appendix A are required by law to be monitored on a frequent schedule (i.e., more than once a year).
• If a specific chemical occurs in drinking water above a set water quality criteria, the public is only warned that the specific chemical is in the drinking water.
• The offending water provider must then increase their monitoring frequency and continue to notify their consumers of the violation for as long as the violation continues.
• No treatment to remove the offending pollutant is required. However, most state water agencies will set a specified time period to reach compliance with drinking water standards. The time to compliance depending upon monitoring data and the reason for the pollution can take years to attain.
• While the violations occur, consumers must either purchase bottled water or treat the water using in-home treatment systems.

An example of these regulatory defects is typified by arsenic pollution in Fallon, Nevada. Fallon became the focus of the U.S. Geological Survey and the national media [11] in April 2001 because of a leukemia cluster that involved 12 children and the occurrence of arsenic in the community drinking water. The concentration of arsenic in the regional groundwater aquifer is approximately 90 ppb, which is almost double the drinking water standard that was in effect at the time. Drinking water is supplied in Fallon by individual groundwater wells and a municipal system that taps this regional groundwater aquifer. Since arsenic was not removed from their drinking water, the community has only been warned of the hazard for the last decade. By the middle of 2003, the City of Fallon will finally begin to treat drinking water to remove arsenic. Unfortunately, Fallon is not the only community that has been affected. According to Congress [12], approximately 15 million Americans are currently exposed to unsafe levels of arsenic. As a result, treatment will be required to reduce their arsenic levels to accepted standards.

The existing monitoring and warning requirements for regulated chemicals in drinking water clearly demonstrate the foolishness of regulatory programs that rely on standards to protect human health. While pollution events can occur at anytime but monitoring only occurs during an extremely short time period, how can the public health be protected? Individuals who can afford to buy an alternate source of drinking water that is not polluted can avoid part of the problem. However, even if a consumer obtains an alternate drinking water source, polluted water for bathing still provides a route of exposure.

Because consumers usually demand a safe source of drinking water, most state regulatory agencies eventually force water utilities to come into compliance with established drinking water standards. The time period required to come into compliance, however, can take years to accomplish if (1) the standard for a chemical of concern is not consistently exceeded or (2) if there is political resistance for immediate change because of economic considerations. Until change comes, the consumer is left to find their own source of safe drinking water. The federal government's claim that USEPA has developed and enforced drinking water standards to protect the public health is the worst form of deceptive advertising.

Although flawed, the water quality standard system has provided monitoring data that can be used to assess the general extent of polluted drinking water in the United States. After collecting years of chemical monitoring data on public water supplies, the USEPA began in August 1999 to assemble this information into a national database. This national database contains the monitoring results for regulated and unregulated chemicals found in drinking water sources that have been distributed by community water systems.

THE NATIONAL DRINKING WATER CONTAMINANT OCCURRENCE DATABASE

As previously noted, not all public water systems are required to monitor for chemical pollutants. The extraordinary cost of chemical analyses also limits the collection of chemical data from smaller community water systems. Therefore, it should be no surprise that of the 168,690 community water systems in the United States only 7.6 percent of these systems have actually monitored for those chemicals with established primary Drinking Water Standards (i.e., the regulated pollutants). For the community water systems that participated in the chemical monitoring, regulated chemical pollutants were detected 19,111 times and in many cases standards were exceeded. These data are summarized in Appendix N. Although the data as summarized by the USEPA database cannot be used to determine the number of community water systems that provided polluted drinking water to their customers, the data suggest that the occurrence of regulated chemicals in drinking water is a common problem. Of even more concern is the occurrence of unregulated chemicals in drinking water.

The USEPA data base also contains monitoring data on 46 unregulated chemical pollutants detected in community water systems. These monitoring data appear in Appendix O. For the community water systems that participated in the chemical monitoring program, unregulated chemical pollutants were detected 5,601 times. Once again, the data as summarized by the USEPA database cannot be used to determine the number of community water systems that provided unregulated pollutants in the drinking water distributed to their customers. More importantly, this database only provides monitoring information on an insignificant number of unregulated chemicals that could potentially pollute drinking water. This information, however, does provoke a more meaningful question. What other unregulated chemicals are also in drinking water? This question can only be answered by even more extensive monitoring programs.

UNREGULATED POLLUTANTS AND MONITORING REGULATIONS

It should be obvious that the occurrence of any known toxic compound in water distributed to consumers should be considered a serious problem. However, the federal government clearly does not consider unregulated chemical pollution of drinking water a concern based on existing unregulated monitoring requirements. For example, in January 2001 the USEPA implemented the Unregulated Contaminant Monitoring Regulations [13] that required 2,800 large public water systems[6] and 800 out of 66,000 small public water systems to conduct assessment monitoring during any contin-

---

[6]CWS and NTNCWS systems that serve more than 10,000 persons. No transient water systems were included.

uous 12-month period[7] for the following chemicals: DCPA mono acid, DCPA di acid, 4,4-DDE, 2,4-dinitrotoluene, 2,6-dinitrotoluene, EPTC, molinate, MTBE, nitrobenzene, terbacil, acetochlor and perchlorate. In addition to these chemicals, the USEPA also required a random selection of 300 large and small public water systems to monitor for the following chemicals: alachlor, diazinon, 2,4-dichlorophenol, 2,4-dinitrophenol, 1-2-diphenylhydrazine, disulfoton, diuron, ESA, fonofos, linuron, 2-methyl-1-phenol, polonium-210, prometon, RDX and 2,4,6-trichlorophenol. If any of these chemicals are detected, each customer will receive a notice each year by July 1, which will identify the chemical pollutants that are in their drinking water. A unique aspect of these regulations is that USEPA is limited to monitoring no more than 30 pollutants in any 5-year monitoring cycle [14].

Given the immense number of potential chemicals that can and do pollute our water resources, this is an absurd monitoring program. Why is the USEPA restricted to 30 chemicals[8]? Because monitoring drinking water quality is expensive. The annual estimated cost just to monitor for the first set of chemicals is $8.4 million. The message is crystal clear: economics is more important than environmental quality. This is no new revelation but just business as usual. Because of the cost, we may never have a comprehensive monitoring program for unregulated pollutants. This fact alone should be justification enough for abolishing the use of drinking water standards as the method by which we protect public health because under the current policies virtually no protection is possible.

Since the USEPA will only be monitoring for a minute fraction of potential pollutants in drinking water, the next issue of concern centers on the probability of the USEPA increasing the number of regulated pollutants in drinking water. Unfortunately, the probability that a substantial number of new chemicals will be added to the Primary Drinking Water List of compounds is highly unlikely.

SETTING NEW DRINKING WATER STANDARDS

Because of the vast number of chemicals that can occur in drinking water, the 1996 Amendments to the Safe Drinking Water Act required the USEPA to develop a list of unregulated pollutants that may pose risks in drinking

---

[7]During this period, quarterly samples must be taken for surface water sources, while only biannual samples are required for groundwater sources.

[8]For example, why aren't many of the unregulated chemicals already identified in our water resources (Appendix K) on this list?

water and to determine which should be added to the Primary Drinking Water Standards. As previously noted, USEPA's consideration of economics will always drive policy decisions that pertain to setting any environmental standard. This reality combined with the enormous number of potential drinking water pollutants has forced the USEPA to identify an economically manageable list of chemicals to determine which these compounds should added to the Primary Drinking Water Standards. This group of chemicals is called the Drinking Water Contaminant Candidate List (CCL).

The first Drinking Water Contaminant Candidate List (CCL) was developed by the USEPA and published in 1998 [15]. This list was developed by the USEPA using the following process. The first CCL included all of the identified toxic or hazardous compounds already regulated by various environmental programs (e.g., Clean Water Act, Resource Conservation and Recovery Act and the Comprehensive Environmental Response, Compensation and Liability Act, etc.). This list contained 391 chemicals. This group of chemicals was then evaluated to estimate which of the compounds had the greatest probability of actually occurring in drinking water. For example, USEPA tried to determine which compounds were produced in the greatest quantities and whether these chemicals would persist in the environment long enough to pollute drinking water resources. Based on these criteria, a final list (listed in Appendix B) of 50 chemicals was selected.

This list of 50 chemicals is an incredibly small number, when one considers that approximately 72,000 chemical substances are listed in the Toxic Substance Control Act inventory of commercial chemicals. Because the USEPA limited its selection process to only those chemicals that were already regulated, the National Research Counsel [16] made the following observation:

This approach, while useful for developing a CCL in a short time period, is like 'looking under the lamp post' because it overlooks potential chemical contaminants not previously identified through inclusion on one of the selected lists. For example, the first CCL development process did not collect and evaluate data on radionuclides, most degradation products of known contaminants, or even all classes of commercial chemicals (such as pharmaceuticals).

Obviously, limiting the selection to only 50 chemicals of concern is not realistic and does the average consumer a great disservice. Using USEPA's own monitoring data for unregulated chemicals as reported in Appendix O, it can be shown that the following chemicals actually occur in drinking water, are known to be toxic, yet are not on the CCL list: carbaryl, cyanazine, aldicarb, aldicarb sulfone, aldicarb sulfoxide, bromochloromethane, bromomethane, butachlor, –butylbenzene, sec-butylbenzene, tert-butylbenzene, o-chlorotoluene, p-chlorotoluene, diacamba, dibromomethane, m-dichlorobenzene, dichlorodifluoromethane, trans-1,3-dichloropropene, 3-hydroxycarbofuran, methomyl, propachlor, n-propylbenzene, toxaphene, 1,2,3-trichlorobenzene, trichlorofluoromehtane, 1,2,3-trichloropropane, trifluralin, 1,2,4-trimethylbenzene and 1,3,5-trimethylbenzene. This example illustrates that the selection of the CCL chemicals based on predefined criteria and economic constraints fails to address the real problem. Drinking water is polluted by known toxic but unregulated chemicals, however, these chemicals are not considered because they do not meet bureaucratic and economic objectives.

For example, when selecting chemicals to regulate, the Safe Drinking Water Act directs the USEPA to identify only those contaminants that may have adverse effect on the health of persons as well as those that pose the greatest public health concerns. Thus, even when a drinking water resource is polluted, only those compounds that are considered to pose the greatest risk can be evaluated. This approach is particularly egregious considering that the true human health effects of many chemicals is largely unknown[9]. In other words, the methods employed by the USEPA to select chemicals for regulation are untimely, incomplete, inconsistent and impractical, and provide just one more important reason why the whole water quality standard system should be scrapped.

In the meantime, the CCL will continue to be modified and updated. The selection process for the CCL is significant because it is from this list that the USEPA must select any chemical that it wants to regulate in the future. More specifically, the 1996 Amendments to the Safe Drinking Water Act require the USEPA within five years after the final CCL is published to make a decision on whether or not to regulate at least five these chemicals. Even with this deadline, it will be years before any new standards are promulgated. This standard setting procedure is discussed in more detail in Exhibit 3.1. Given the level of pollution that exists in our nation's surface

---

[9]For example, the USEPA's existing chemical toxicity testing and assessment program is only focusing on the approximately 3,000 high production volume chemicals (i.e., those produced at levels over 1 million ponds per year). The USEPA has a long way to go considering only 550 of these chemicals have been evaluated since 1979 [16].

Exhibit 3.1   You Do Know That Science Has Nothing To Do With Policy—How Drinking Water Standards Are Set In The United States

According to the USEPA*, the United States enjoys one of the best water supplies in the world. The United States has also spent hundreds of billions of dollars to build drinking water treatment and distribution systems and currently spends $22 billion per year to operate and maintain that system. Why is it necessary to spend these large amounts on drinking water? USEPA believes that "all sources of drinking water contain some naturally occurring contaminants."

The USEPA controls the levels of contaminants that are allowed in public drinking water systems by setting National Primary Drinking Water Regulations (NPDWR). Drinking water standards, or MCLs, have been set for more than 80 contaminants. MCLs are established based on known or potential health effects, the availability of technologies to remove the contaminant, their effectiveness, and the cost of treatment. MCLs are supposed to be set at levels that protect public health. State public health and environmental agencies have the primary responsibility for ensuring that these federal drinking water quality standards, or more stringent ones required by an individual state, are met by each public water supplier.

Prior to 1974 each state set the standards that had to be met at the local level. Drinking water quality and protection standards differed from state to state. In 1974, the Safe Drinking Water Act (SDWA) was passed to ensure that a consistent level of protection and quality existed throughout the United States. As part of the 1996 SDWA amendments, USEPA is now required to publish a list of contaminants that are not currently subject to a NPDWR and are "known or anticipated" to occur in public water systems. This list, the Contaminant Candidate List, is supposed to set research priorities, help in the development of guidance from USEPA and the selection of contaminants for making regulatory determinations and/or monitoring by the States.

The CCL currently consists of 50 chemical and 10 microbiological contaminants. Contaminants on the list include industrial solvents, metals, pesticides, explosives, rocket fuels, biocides, and common elements.

*USEPA (2001). "Water on Tap: A Consumer's Guide to the Nation's Drinking Water" URL: http://www.epa.gov/safewater/wot/introtap.html.

(Exhibit 3.1, continued)
Exposure to the listed chemicals is known to lead to a host of significant health effects including cardiovascular, pulmonary, immunological, neurological, and endocrine (e.g., diabetes) effects, cancer and even death. The SDWA required USEPA to determine whether or not to regulate not less than 5 of the contaminants (but not all of the contaminants) from the CCL by 2001. The USEPA must revise the CCL by February 2003 and a decision inclusion of any of the chemicals on the second CCL can be deferred until August 2006, eight years after the first CCL was created. If a chemical that is deferred until 2006 for regulation, then a proposed NPDWR for that chemical must be issued no later than August 2008. The final rule for any chemical on the CCL is required 18 months later (February 2010) and can be delayed an additional nine months (November 2010). Under the SDWA, water systems will have three years to comply with an NPDWR and may take an additional two years if capital improvements to the water system are necessary. Therefore, it is possible for a chemical that was recognized by USEPA in 1998 as a potential health threat will not be removed from a drinking water system until 2013! After spending up to 15 years evaluating the potential threat from a contaminant and making improvements to aging water distribution systems, the public may finally get relief from exposure to that contaminant!

The delay in the decision making process is understandable when that time is spent gathering critical data on the potential health effects of exposure to contaminants. What is not understandable is the amount of time, money, energy spent by various advocacy groups to delay the decision making process. Consider a set of recent decisions by USEPA on the proposed MCL for arsenic, a known human carcinogen. On January 22,

and groundwaters, this rate of regulation is clearly unacceptable. Thus, the current water quality policies, as implemented by the USEPA cannot protect the public health.

A FAILED POLICY

The monitoring data for both regulated and unregulated chemicals, which cover a total of 112 compounds (Appendix N and O), suggest that thousands of community water supplies provide chemically polluted water to their consumers. Sadly, the exact number of Americans who drink or bath in this polluted water is unknown. These monitoring data raise the following questions:

(Exhibit 3.1, continued)
2001, USEPA published a "final" ruling on a proposed MCL for arsenic, reducing it from 50 micrograms per liter (ug/L) to 10 ug/L (*Federal Register*, 66 FR 6976). This drinking water standard was based on the standard set by the United States Public Health Service in 1943. The current World Health Organization (WHO) standard is 10 ug/L. USEPA staff was interested in setting the standard at 5 ug/L, but changed it to 10 ug/L following intense criticism that USEPA's cost-benefit ratio was unrealistic and their estimated cost for mitigation was excessively low. The "final" rule was to become effective March 23, 2001 and compliance with the rule required by January 2006. On March 20, 2001, USEPA withdrew the January 22, 2001 standard for arsenic and sought an independent review of the science behind the standard and the cost estimates to communities that would have to implement the rule. On April 23, 2001, USEPA announced its intention to put in place an arsenic MCL and would require compliance with that standard by 2006. USEPA also asked the National Academy of Science (NAS) to review a range of 3 to 20 ug/L for the new drinking water standard.

While the USEPA spent considerable time and effort developing a proposed MCL for arsenic, a known human carcinogen, the MCL was tabled not by concerns over the science but rather over the cost of implementing that standard! As a result, the prospect for developing MCLs for the 50 contaminants on the CCL in a timely manner are limited at best. As an attorney for an industrial advocacy group against the new standard pointed out "science has nothing to do with policy."

- What other chemicals are actually present in drinking water since only 112 of the tens of thousands of chemicals used in the United Stated are monitored?
- What chemicals occur in the water supplied from the approximately 140,000 community water systems that do not monitor their product?
- For the regulated and unregulated chemicals that are monitored but reported as not detected, what is their actual concentration in drinking water? Just because standard methods of analysis do not detect a specific chemical does not mean that the chemical is not present. This simply means that more sensitive methods of analysis need to be used to lower the range of detection.

• What chemicals actually occur in drinking water during those time periods when drinking water is not monitored (i.e., more than 99 percent of the time)?

The fact that these questions have not been answered demonstrates the failure of the chemical monitoring to supply meaningful data to the American consumer. Other than by warning the public not to use polluted water resources, this failure also suggests that there is no current safeguard for protecting the American public against drinking chemically polluted water, bathing in chemically polluted water or inhaling volatile organic pollutants from chemically polluted water. Such warnings, however, assume that a community water system, a private water distributor or a private well owner knows that their water is chemically polluted. This assumption is neither accurate or realistic.

Furthermore, it is impossible to monitor for all potential chemical pollutants in drinking water on a national level because 1) the laboratory performing the analysis needs to know what specific chemicals to look for in the water, 2) most laboratories require standard methods of analysis for each specific chemical, but these methods are not routinely available and 3) not be enough money is available to pay for this type of program. Because this is an impossible task, the extent of the existing chemical pollution in drinking water cannot be monitored. Without such monitoring, the American public will continue to unknowingly consume chemical pollutants. Based on these conditions, it is reasonable to conclude that the current regulatory system of standard based pollution, founded on chemical monitoring and warnings, is a failure.

THE NEED FOR PRECAUTION

Pollution of our nation's water resources is sanctioned by both state and federal governments. Our reliance upon theories of risk, minimum water treatment to meet water quality standards, and monitoring only for those chemicals of concern to state and federal agencies are the only safeguards protecting our health. This system of providing "safe" drinking water to the American public, however, provides only the illusion of safety. As a result, it is reasonable that individuals exercise some degree of precaution before blindly accepting the safety of the water they consume. Many consumers have been doing so by drinking bottled water. A survey [17] conducted by the American Water Works Association in 1993 reported that 35 percent of people were worried about tap water safety and that over half of all Americans drink bottled water. Another study [18] of the risk perception that drives bottled water use found that among other reasons "public perceptions of chemical risk are also influenced by the level of distrust the public holds for government and industry." This perception could be one of the reasons that sales of bottled water rose to $5.2 billion in 1999 and are expected to increase by a compound rate of 15 percent over the next five years [19]. In terms of consumption, this equates to approximately 3.4 billion gallons annually or over 12 gallons per person. These data suggest that consumers believe that drinking bottled water is one solution to reducing the consumption of chemical pollutants.

Bottled Water

In general, the quality of bottled water is expected to be better than what comes out of the tap. After all, the product is in a bottle and is expensive relative to tap water. Because of this perception, providers of bottled water should be held to a higher standard of purity. Unfortunately, they are not. In fact, bottled water is not necessarily a better than tap water and can be just as polluted as tap water. Furthermore, bottled water is not regulated by the USEPA but rather by the Food and Drug Administration (FDA) under Title 21, Part 165.110 of the Federal Code of Regulations[10]. Under these regulations, bottled water must comply with the National Primary and Secondary Water Quality Standards[11] (see Appendix A). The major difference between USEPA and FDA oversight lies in biological purity standards.

For example, the standards for biological purity required by the USEPA and the FDA are compared in Table 3.1. This comparison shows that USEPA standards are more comprehensive than FDA standards[12]. Even with this dichotomy, the good news concerning bottled water is that for the past 37 years there have been no confirmed reports in the United States of illnesses or disease linked to bottled water. However, the potential for microbial pollution of bottled water still exists.

The bad news is that a recent survey conducted over a four year period by the Natural Resources Defense Counsel [20] found that one third of the 103 brands of bottled water tested contained elevated levels of bacteria, inorganic chemicals and/or organic chemicals. In their study, the specific chemicals that were tested for varied widely between laboratories. However, in gener-

---

[10]Some states have standards that are stricter than the federal standards.

[11]Chemical monitoring frequency is essentially the same between the USEPA and FDA neither of which are adequate.

[12]Since bottled water is classified as a food by the FDA, why do FDA regulations on pesticides (see Appendix E) in food not apply to water? For some reason (probably economic) the FDA believes that chemicals, which are banned from foods, are not found in water.

Table 3.1
**Comparison of Biological Purity Standards
Between the USEPA and FDA for Bottle Water**

Parameter	USEPA *Tap Water*	FDA *Bottled Water*
Disinfection Required	Yes	No
E-Coli & Fecal Coliform Banned	Yes	No
Testing Frequency for Bacteria	Hundreds/month	Once/week
Filtration to Remove Pathogens or Have a Protected Source	Yes	No
Must Test for Viruses	Yes	No

al the chemicals looked for were those with established primary drinking water standards. The chemicals found in bottled water include acetone, n-butylbenzene, 2-chlorotoluene, dichloroethane, ethylbenzene, p-isopropyltoluene, methylene chloride, styrene, trichloroethylene, toluene and xylene. This list of chemicals is consistent with those compounds found in drinking water sources throughout the United States (see Appendix N and O). As a result of these data, the Natural Resources Defense Counsel report concluded that bottled water was not necessarily safer than tap water[13].

Bottled water can also contain disinfection byproducts if it comes from community water systems or the bottle water company uses chlorine or bromine compounds for disinfection. In either case, this water could contain any number of the TTHM and HAA5 compounds. Since the Stage 2 Rule will amend the Primary Drinking Water Standards for the TTHM and HAA5 compounds, bottled water must ultimately meet the same standards (*Federal Register*, March 28, 2001).

The degree of safety offered by bottled water depends on the "source water" used by the bottler and on whether that water is purified to remove chemical pollutants. The different types of bottled water products are listed in Table 3.2. Unfortunately, none of these descriptive terms really provide a guidance as to the environmental characteristics or quality of the water. However, many bottlers do treat their water to remove chemical pollutants. For example, both Aquafina and Dasani purify their products using

---

[13]Both government and industry estimate that between 25 and as high as 40 percent of all bottled water is actually tap water.

Table 3.2
Different Types of Bottled Water

Water Product	Source Water
Artesian Water	Well water from a confined aquifer
Mineral Water	Water containing not less than 250 parts per million as total dissolved solids
Spring Water	Water derived from an underground formation from which water flows naturally to the surface of the earth
Groundwater Well	Water from a hole bored, drilled or otherwise constructed in the ground and tapping into an aquifer
Sparkling Water	Water that has some residual amount of carbon dioxide which was in the water at its source
Purified Water	Water that has been produced by distillation, deionization, reverse osmosis or other process (sometimes called distilled or deionized water)

Notes:
1. Label statement: Any water that comes from a community water system must be labeled as such with a label that states: "from a community water system" or "from a municipal source."
2. "Mineral" and "sparkling" waters usually originate from artesian wells, groundwater wells or springs.

reverse osmosis technologies. Aquafina advertises that there is "nothing" in their product. After removing a significant portion of dissolved chemicals and minerals, Dasani puts minerals back into their product to enhance the taste. One of the most unique products on the market is Penta which is advertised as being purified, fluoride free and MTBE free.

In addition to the chemical pollutants that can be present in the natural bottled water, the consumer should also be aware of several issues associated with the plastic bottles in which it is often sold. The plastic of choice by bottled water companies is poly(ethylene terephthalate) or PET. This plastic is recycled and recycled PET can contain organic pollutants from previous uses [21]. In the absence of specific proof that PET bottled water products do not contribute trace chemical pollutants to their products[14], the

[14]FDA regulations require that these plastics do not produce a chloroform-soluble extract value greater than 0.5 mg/in^2 for distilled water with a contact time of 2 hours at 250 F^0.

prudent consumer would purchase either water in glass bottles or if plastic containers cannot be avoided, use a disposal carbon filter that is specifically designed for plastic bottles[15].

Here are some general guidelines for selecting a bottled water whose product has the least probability of containing chemical pollutants.

- Do not drink bottled water whose source is from a "community" or "municipal" water utility that has not be purified.
- Do not drink bottled water whose source is from an artesian well, spring, or groundwater well unless it can be determined from the provider that the source water originates from a deep aquifer and that the landscape of the recharge zone[16] is dominated by wilderness or a land use having a minimum of human development. An example of a water source that originates from a deep aquifer and wilderness landscape would be the bottled water sold by Mountain Valley in Arkansas.
- Regardless of the source of bottled water, all bottled water should be labeled to indicate that their "chemical purity" for man-made chemicals is unknown[17]. The exception to this rule would be for bottled water that is treated using the best available technology or combination of technologies to reach a level as close to "zero pollution" as possible. It should also be recognized that using this level of treatment will remove all naturally occurring minerals as well as chemical pollutants. Thus, those individuals concerned about electrolyte balance (e.g., athletes) should be aware that such water will lack such minerals.

Drinking bottled water that is chemically pure is one step towards limiting a person's exposure to pollutants[18]. However, for those individuals who wish to minimize their exposure and their family's exposure to chemical pollutants in water, the only real alternative is to treat the water prior to its consumption in the home. This precaution is further justified given the potential for the intentional poisoning of a water resource as the result of a terrorist act.

---

[15]If a carbon filter is used, the consumer should make sure that the manufacturer's recommendations on the volume of water that can be treated are followed.

[16]The area at the earth's surface where surface water (i.e., rivers, lakes, rainfall) infiltrates through the soil and percolates downward until it flows into or "recharges" an aquifer. The longer the travel time of the water (i.e., the time between recharge and the water's exit at the spring) the better.

[17]In reality, you can't test for all the potential chemicals that can occur in drinking water.

[18]Having this level of purity in bottled water then raises the questions as to what is the quality of water used in soft drinks and beer? Obviously, if there are chemical pollutants in bottled water, why not in soft drinks, reconstituted juices and beer?

POTENTIAL POLLUTION BY TERRORISTS

A USEPA sponsored study was conducted in the early 1970s [22] to address the threat of the intentional poisoning of drinking water resources. This study concluded that the release of a chemical or biologic agent into a drinking water resource could occur in two fundamental ways. Hazardous materials could be either (1) introduced into a water resource reservoir prior to its treatment and release into the water distribution system or (2) injected into drinking water that is already within the water distribution system pipelines (i.e., post-treatment). These same conclusions were expressed by the American Water Works Association in February 2002 [23].

The pre-treatment threat was considered not to be very great. This finding was based on the fact that large community reservoirs generally contain tens of millions to billions of gallons of water. Such a enormous volume of water would dilute any toxic chemical or biologic agent to the point of being ineffective. In order to create a hazardous condition, the perpetrator would need to dump truck loads of chemicals or hundreds of pounds of biologic agents. However unlikely, it is not impossible to obtain truck loads of toxic chemicals (i.e., as a licensed transporter or by high-jacking) in various location of the United States. Still, it would very difficult for terrorists to produce, steal or smuggle hundreds of pounds of biologic agents into the United States for the purpose of polluting water supplies. Furthermore, once the water was chlorinated at the water treatment plant most biologic agents would be destroyed.

These conclusions were basically supported by an 2001 article in *Water Environment and Technology* [24] However, this article did point out that with over 6,800 public drinking water intakes[19] on rivers in the United States, these intakes "can be considered vulnerable to disruption by accidental or intentional release of hazardous chemicals or biological substances." Given the potential for the intentional pollution of water resources, Congressional hearings in November 2001 on antiterrorism [25] concluded that existing water treatment technologies could be put together in a series to provide "multiple barriers" to block the chemical or biological pollution. In other words, the technical solutions exist, they only need to be implemented. Still, it was felt that an intentional act of polluting water resources prior to treatment was a serious threat.

The greatest threat to post-treatment water supplies would be from toxic chemicals introduced into small water storage tanks. Doing so would be a

---

[19]This number of intakes on rivers also indicates the magnitude of public water supplies that can be impacted by upstream pollution without any accident or intentional act.

difficult task considering that the terrorist would have to be familiar with the storage system, be able to gain access to the area and remain undetected during the time necessary to completely pump the chemicals into the tank. The earlier USEPA study also suggests that the post-treatment release of toxic chemicals into the water distribution system could constitute a serious problem. On one hand, the introduction of a biologic agent into a post-treatment water supply line is much less of a hazard since residual chlorine or bromine usually present (unless ultra violet light is being used)[20]. The real threat exists from introduction of a toxic chemical into the distribution system outside the confines of the treatment plant. In many cities, this aim can be accomplished by simply pumping a truck load of a toxic chemical into any fire hydrant. The potential number of individual households or businesses that could be effected would obviously depend upon their proximity to the point of injection. These same concerns were mentioned in the November 2001 article by the American Water Works Association [26].

On October 10, 2001 the Federal Bureau of Investigation (FBI) provided a statement for the Congressional Record on "Terrorism: Are American's Water Resources and Environment at Risk." The conclusions of the FBI report were generally the same as the 1972 study. Among their conclusions were that (1) pollution of a water supply with a biologic agent that causes illness or death is possible but not probable, (2) pollution of a water resource with a biological agent would unlikely to produce a large risk to public health, (3) a successful attack would require knowledge of the water supply system and (4) a successful attack would likely involve a post-treatment injection. As a result, the FBI recommended that water utilities "maintain a secure perimeter around the source (if possible) and the treatment facility. In addition, security should be maintained around critical nodes such as tunnels, pumping facilities, storage facilities, and the network of water mains and subsidiary pipes should be enhanced."

Although the FBI recommended increased security on the post-treatment distribution system, there were no specific recommendations for locked or secured water mains and fire hydrants or electronic monitoring. Given the cost associated with such a program, along with the FBI's estimate of the low probability of a successful attack or potential for large scale damage to a significant number of consumers, it is highly unlikely

---

[20]Some community water systems are switching to ultraviolet light for disinfection to avoid halogenated disinfection byproducts. Thus, these distribution systems would be much more vulnerable to biological attack.

that federal funding would be available for such security measures. In addition to the FBI's assessment, the American Water Works Association issued a press release on October 18, 2001 that proclaimed "Terrorist Threats to Nations's Drinking Water Supply Remote."

These warnings are still important because they confirm that a threat to our drinking water from a terrorist act is possible. No matter how remote this threat may be it should be remembered that such an attack does not require sophisticated technology or a lot of money. Furthermore, the safety of drinking water cannot be guaranteed because (1) there are multiple points of attack in a post-treatment distribution system and (2) these systems require extensive and costly monitoring to detect any breach in security.

Given the cost and the dispersed nature of the problem, it is not very likely that a comprehensive security program will be established. This reality, when combined with the existing pollution of our drinking water, suggests that in order to protect the public a national program should be adopted that will implement comprehensive in-home and workplace water treatment.

THE NEED FOR INCREASED PROTECTION

A significant portion of this nation's drinking water is polluted with both inorganic and organic chemicals. These water resources cannot be shown to be "safe" by compliance with federal or state water quality standards. Over the next decade, we can expect a marked improvement in drinking water quality. This improvement will not be because of additional chemicals being added to the Primary Drinking Water Standards, but because community water systems will be forced to install advanced water treatment technologies in order to remove disinfection byproducts. These technologies will not only remove disinfection byproducts, they will also remove a large number of other chemical pollutants.

Given that many community water systems will be adding advanced treatment technologies to remove disinfection byproducts, they should complete the process by adding the best available technologies with the goal of achieving a water product with a minimum concentration of chemical pollutants. Even without the new Stage 2 Rule on disinfection byproducts, all community water systems should have this same goal. The health of the American public cannot be protected by simply drinking bottled water. Instead, there needs to be a fundamental change from standard based water quality to technology based water quality.

The necessary water treatment technologies exist today for both community water systems and consumers served by private water sources to treat their water to produce a product with a minimum concentration of chemical pollutants. No technical barriers prevent us from minimizing pollution, only political and economic obstacles must be addressed. Even if a technology based drinking water policy were adopted today, it will take decades for communities and private water sources to implement such a policy. Until community water systems adopt a minimum pollution policy, water utility consumers and those with private water sources do have another option. This option is in-home and workplace water treatment systems.

IN-HOME AND WORKPLACE WATER TREATMENT

For those consumers who want a source of drinking water with lower levels of chemical pollution and also want to protect their family from intentional chemical or biologic attack, there is only one real option. Treat the water you consume[21]. Such precaution can be expensive. Yet, many households in the United States choose to treat their water to remove pollutants or improve the taste and odor. Annual sales of home water treatment systems have been reported [27] to have probably exceeded 6 million units annually as of 2001[22].

Although there are various methods of removing pollutants from water, the most commonly used techniques employed by in-home water treatment systems are reverse osmosis (RO) and granulated activated carbon (GAC). RO technologies use fine porous membranes to separate inorganic and organic chemicals from water. These units are effective at removing dissolved salts, suspended matter, and a wide variety of dissolved organic chemicals as well as bacteria and viruses. A typical "claim list" of chemicals removed by an RO system is given in Table 3.3, while Table 3.4 provides a example of RO removal efficiencies. Carbon filters (GAC) can reduce chlorine, many man-made synthetic chemicals (chlorinated and non-chlorinated) including pesticides, some radiological constituents, fluoride, radon and some metals [28]. An example of removal efficiencies using GAC is given for a set of selected chemicals in Table 3.5.

As can be seen from Tables 3.3 through 3.5, a significant reductions in chemical pollutant levels can be achieved using these technologies. However, chemical removal efficiencies are only known for a small number of compounds (i.e., the regulated chemicals). Therefore, it must be

---

[21]This means the water you drink, bath in or inhale (i.e., water vapor and volatile organics from showers, dish washers and laundry).

[22]Sales were at approximately 3.4 million units in 1991 and reached 4.7 million units in 1996.

Table 3.3
Reverse Osmosis Chemical Reduction Claim

alachlor	1,2-dichloropropane	styrene
atrazine	cis-1,3-dichloropropylene	1,1,2,2-tetrachloroethane
benzene	dinoseb	tetrachloroethane
carbofuran	endrin	toluene
carbon tetrachloride	ethylbenzene	1,2,4-trichlorobenzene
chlorobenzene	ethylene dibromide	EDB
dibromochloropropane	1,1,1-trichloroethane	DBCP
heptachlor	1,1,2-trichloroethane	o-dichlorobenzene
p-dichlorobenzene	heptachlor epoxide	trichloroethylene
hexachlorobutadiene	trihalomethanes	xylenes (total)
hexachlorocyclopentadiene	1,2-dichloroethane	lindane
trans-1,2-dichloroethylene	methoxychlor	1,1-dichloroethylene
dichloroethylene	2-4-D	Cis-1,2-dichloroethylene
pentachlorophenol	2,4,5-TP (silvex)	simazine
total dissolved solids	barium	cadmium
copper	hexavalent chromium	trivalent chromium
lead	radium	selenium

Source: Eco Home Products Hydro Line 5000 Reverse Osmosis Unit, 7745 Alabama Ave. #11, Canoga Park, CA 91304).

Table 3.4
Reverse Osmosis Chemical Removal Efficiencies

Compound	Percent	Compound	Percent
Aluminum	97–98	Polyphosphate	98–99
Bromide	93–96	Pyroge	99+
Cadmium	96–98	Radioactivity	95–98
Chloride	94–95	Silica	85–95
Chromate	90–98	Silicat	95–97
Chromium	96–98	Silver	95–97
Copper	98–99	Sodium	94–98
Cyanide	90–95	Strontium	96–99
Ferrocyanide	99+	Sulfate	99+
Hardness	95–98	Thiosulfate	99+
Iron	98–99	Virus	99+
Lead	96–98	Magnesium	96–98
Ammonium	85–95	Manganese	98–99
Arsenic	94–96	Mercury	96–98
Bacteria	99+	Nickel	98–99
Barium	96–98	Nitrate	93–96
Bicarbonate	95–96	Orthophosphate	98–99
Borate	40–70	Phosphate	99+
Boron	60–70		

Source: The Good Water Company, 151 N. Main Street, Suite 700, Wichita, KS 67202.

assumed that other unregulated compounds will also be removed with the same level of efficiency. This assumption is generally valid but not without some degree of risk. For example, compounds such as carbon disulfide, methyl bromide, choromethane and dichlorodifluoromethane (freon 12) would not be substantially removed using GAC. Without actual removal efficiencies for a vast number of unregulated compounds, it will be necessary to combine both RO and GAC into one treatment system in order to provide a level of treatment that ensures the greatest level of pollutant removal.

The amount of water treatment required largely depends upon the source of a home's water supply. Water from a community water supply will

Table 3.5

Activated Carbon Performance Test

Chemical	Removal Efficiency (Percent)
Chlorine	98.0
Dichloromethane	98.9
Chloroform	99.5
Trichlorethylene	99.1
Perchlorethylene	99.6
Benzene	99.3
Toluene	99.3
p-Xylene	99.4
Aldrin	98.5
HCH	99.4
p-DDT	98.0
PCB	95.0
Atrazine	99.0
Phenols	99.3
Napthalene	99.2
Fluoranthene	98.4
Benzo-a-pyrene	94.7

Source: Katadyn Water Filters (Katadyn can be contacted at webmaster@Katadyn).

require less treatment than water from a private well. In other words, the greater the amount of inorganic and organic compounds that are in the source water, the more frequent the maintenance (i.e., changeout of RO membranes and/or GAC filter media). As a result, some waters that contain elevated concentrations of calcium and magnesium (i.e. hard water) should be softened[23] to reduce potential scaling of the treatment system. Iron can also reduce the efficiency of RO membranes. As a consequence, it is recommended that if iron levels in the source water are above 5 parts-

---

[23]This is usually accomplished using ion exchange technologies where sodium is exchanged for the calcium and magnesium in the water (sodium carbonates do not form solids like calcium and magnesium carbonates).

per-million, they should be removed prior to an RO system in order to reduce RO maintenance.

The maintenance frequency of any water treatment system is ultimately depends upon when pollutant pass-through or breakthrough occurs. At this point a new RO membrane or fresh GAC filter media should be installed. How does a home owner determine when pollutants are no longer being removed? This calculation is fairly straight forward for RO technologies. When an RO membrane begins to clog, the pressure difference across the membrane will increase. This pressure difference can be monitored electronically to warn the home owner that it is time to change the membrane. In addition, when the efficiency of an RO membrane decreases, the salt content of the treated water will increase. Salt levels can also be monitored using a simple salinity probe. The problem with GAC filter media is that there is no way to determine when breakthrough of organic pollutants occurs. The only way to determine if low concentrations of a compound are no longer being adsorbed by GAC is by analyzing the water for the occurrence of that compound. Given the wide range of compounds that can occur in drinking water, this option is not a realistic monitoring method. Thus, GAC filter media should be changed out on a pre-defined maintenance schedule designed to ensure that sufficient capacity to remove organic pollutants remains[24]. This changeout cycle should be determined on the basis of the amount of organics that will pass through the GAC filter for a given volume of treated water. This filter cycle can be determined in consultation with the filter manufacturer or the filter supplier.

When a home owner uses a home treatment system, he or she must assume responsibility for maintaining it. Failure to do so will result in either a decreased pollutant removal efficiency or no pollutant removal at all. If the homeowner does not want to assume these *very important* responsibilities, then they should engage a water purification service company to provide routine monitoring surveillance and maintenance activities.

Home purification systems fall into two broad categories, point-of-use systems and point-of-entry systems. A point-of-use system is installed at the location of the water's use, while a point-of-entry system is installed to treat all water entering the house. Since point-of-use systems only provide a means of suppling chemically pure water to one location in a household, they are not universally recommended as they do not treat bath water, dishwasher water or water used for the washing machine. However, if space

---

[24]GAC can also be an excellent place for bacteria to grow. Thus, GAC should be changed out on a regular basis to minimize bacteria pollution.

limitations or economic considerations do not allow the use of a point-of-entry system, a point-of-use system can be used.

In most homes, point-of-use systems are usually installed at the kitchen sink. The two most common point-of-use systems are GAC filters attached to water tap and under-the-sink units that use RO, GAC, or a combination of both methods. Of these two methods, GAC filters on the tap, although commonly used, are not recommended because under normal household use, water may not come in contact with the GAC filter for a long enough time period to effectively remove pollutants. Furthermore, it is almost impossible to keep track of how much water has passed through the filter[25]. As a result, it is highly likely that the adsorbing capacity of the GAC filter will be exhausted (i.e., no longer providing any treatment) without the homeowner's knowledge.

Furthermore, potential for bacterial growth in both RO and GAC units can become a serious issue. These treatment units can serve as a potential location for microbial growth. As a result, all under-the-sink treatment systems should include a ultraviolet light unit to destroy microorganisms. In addition to this problem, if the source of the drinking water has hard water (i.e., a high levels of calcium and magnesium) or contains more than 3 parts-per-million iron, it should be treated at the point-of-entry to remove these materials. Pretreatment for hardness and iron removal will increase treatment efficiency and reduce maintenance.

A good point-of-use system at the kitchen sink should include (1) a pre-filter to remove turbidity or suspended solids and chlorine, (2) a canister system employing a high flow OR unit followed by an GAC unit, and (3) an ultraviolet light treatment unit. These systems will require a four to five gallon storage tank between the GAC unit and the ultraviolet light treatment unit. If the size of the system does not fit under the sink, it can be placed in either a basement or garage with the appropriate plumbing to the point of use.

A point-of-entry system takes a portion of the total amount of water that enters the home, treats it and stores it for future distribution to all sinks, showers, bath tubs, and household appliances. The remaining untreated water[26] can then be distributed to outside faucets and irrigation systems. Generally, a point-of-entry system has the same equipment as a point-of-use system except the individual components are larger and more expensive. For example, a minimum storage tank is usually 100 to 300 gallons

[25]A filter can only treat a specific volume of water—for example, 60 gallons.
[26]Which is most of the water used by a household.

and should be glass lined. These systems may require as much as a 50 to 70 square foot space in a garage, basement or utility room.

What any point-of-use or point-of-entry system can accomplish is easily summarized by reviewing the respective performance characteristics of both RO and GAC components (see Tables 3.3, 3.4 and 3.5). The pre-filter removes turbidity and also helps to delay early fowling of the RO unit. Such filtration is especially important if the source water is not from a community water supply. The RO system removes a wide variety of chemical and biological pollutants. Removal efficiencies approaching 95 percent are routinely obtained with some chemicals having a 99 percent removal rate. The GAC unit acts as a polishing step, removing constituents which might have passed the RO unit. The GAC unit also operates at efficiencies typically exceeding 90 percent. Thus, using both RO and GAC units in combination should result in excellent water quality. Because the GAC is uses as a polishing step, probability of any organic chemical passing through the treatment system is lower when the GAC filter is replaced on a routine basis. GAC units used in this manner are also less likely to have a bacterial problem since level of organic compounds, suspended sediments and nutrients in the filter media are generally low. The UV unit assures disinfection of any bacterial pollutants that may have passed through the previous units. It should also be noted that UV lamps must be replaced once or twice a year and that water storage tanks should be routinely cleaned usually on an annual basis.

Because both RO and GAC treatment units have unknown removal efficiencies for unregulated chemicals it is critical that they be combined in the same treatment system. This issue aside, it is widely accepted that these treatment units when properly maintained will achieve overall removal efficiencies of 90 to 95 percent for dissolved inorganic and organic compounds. At present, in-home treatment systems are not regulated by federal, state or local laws. However, the industry is "self-policed" by several organizations including the National Sanitation Foundation (NSF). The NSF, a not-for-profit group, has been accredited by the American National Standards Institute, the Occupational Safety and Health Administration and the Standard Council of Canada. NSF's program certifies the performance of in-home water treatment units and system components. The NSF does not certify complete systems. It is highly recommended that only NSF certified treatment units and components be used for in-home treatment systems.

Although water treatment systems require routine maintenance in order to guarantee water quality, this fact should not negate the value of their use.

Furthermore, for those individuals that are unsure how to maintain a water treatment system or just do not want to be bothered with having to remember maintenance cycles, a service company can be hired to monitor, clean and replace filters as necessary. Ultimately, as more consumers install in-home treatment systems, reductions in cost of these treatment technologies and their maintenance will be probably realized.

References

1. Sean Gray, et al, "Consider the Source, farm runoff, chlorination byproducts, and human health," *Environmental Working Group*, Washington, DC (January 2002).

2. Roberts, Megan G., Philip C. Singer and Alexa Obolensky, "Comparing Total HHA and Total THM concentrations Using ICR Data," *Journal of the American Water Works Association* (January, 2002).

3. Najm, Issam and R. Rhodes Trussell, "NDMA Formation in Water and Wastewater," American Water Works Association, Water Quality Technology Conference, Salt Lake City (November, 2000).

4. USPHS, "9th Report on Carcinogens," U.S. Department of Health and Human Services, National Toxicology Program, Washington DC (January 2001).

5. Stecher, P. G., M. J. Finkel, O. H. Siegmund, and B. M. Szafranski (editors), *The Merck Index of Chemicals and Drugs*, 7th Edition, Merck & Company, Rahway, New Jersey (1960).

6. Harris, Robert H. and Edward M. Brecher, "Is The Waster Safe to Drink, Part 1: The Problem," *Consumer Reports* (June 1974).

7. Liang, Sun, et al., "Treatability of MTBE-contaminated groundwater by ozone and peroxone," *Journal of the American Water Works Association* (June 2001).

8. Federal Water Review (Sept./Oct. 2000).

9. Personal communication with Jerry Gilbert (see Forward).

10. USEPA, "Providing Safe Drinking Water in America" Office of Enforcement and Compliance Assurance, Washington, DC, EPA 305-R-00-002 (April 2000).

11. USGS, "The Environment and Human Health," U.S. Geological Survey, Fact Sheet FS-054-01 (May 2001).

12. Report to Rep. Henry A. Waxman, "Public Exposure to Arsenic in Drinking Water," Special Investigations Division, Committee on Government Reform for the U.S. House of Representatives (October 4, 2000).

13. USEPA, "Unregulated Contaminant Monitoring Regulation: Monitoring for List 1 Contaminants by Large Public Water Systems, Office of Water, EPA 815-F-01-003 (January 2001).

14. USEPA, "Final Revisions to the Unregulated Contaminant Monitoring Regulation," Office of Water, EPA 815-F-99-005 (August 1999).

15. USEPA, "Announcement of the Drinking Water Contaminant Candidate List; Notice," *Federal Register* 63 (40): 10274–10287.

16. National Research Council, *Identifying Future Drinking Water Contaminants*, National Academy Press, Washington DC (1999).

17. American Water Works Association Research Foundation, "Consumer Attitude Survey on Water Quality Issues," (1993).

18. Anadu, E. C. and A. K. Harding, "Risk Perception and Bottled Water Use," *Journal of the American Water Works Association* (November 2000).

19. Bottled Water Web (http://www.bottledwaterweb) (3/28/2001).

20. Natural Resources Defense Council, "Bottled Water, Pure Drink or Pure Hype?" (March 1999).

21. Komolprasert, V., et al., " Volatile and Nonvolatile Compounds in Irradiated Semi-rigid Crystalline Poly(ethylene terephthalate) Polymers," *Food Additives and Contaminants*, Vol. 18 (2001).

22. Agardy, Franklin J., "The Threat . . . from Additions of Chemicals and Biologicals to a Municipal Water Supply," URS Research Company (May 1972).

23. Regush, Nicholas, "Questions on Protecting US Water Supplies," *Journal of the American Water Works Association* (February 2002).

24. Anonymous, "WE&T News Watch, Water-related Bioterrorism Unlikely, Experts Say," *Water Environment and Technology*, Vol.13 (December 2001).

25. Luthy, Richard, G., "Safety of Our Nation's Water," Statement before the Committee on Science, U.S. House of Representatives Hearing on: H.R. 3178 and the Development of Anti-Terrorism tools for Water Infrastructure (November 14, 2001).

26. Anonymous, "Manager to Manager, Are We Prepared?" *Journal of the American Water Works Association*, Vol. 93 (November 2001).

27. Michigan State University Extension, Water Quality Bulletins No. WQ239201 (July 1997).

28. Seelig, B., F. Bergsrud and R. Derickson, "Treatment Systems for Household Water Supplies—Activated Carbon Filtration," North Dakota State University, NDSU Extension Service, AE-1029 (February 1992).

Chapter 4

Policy

"How clean is *clean enough* can only be answered in terms of how much we are willing to pay and how soon we seek success."
— Richard Nixon, Council on Environmental Quality, 1971.

Because the public cares about the quality of the water they drink, every state in the nation enforces at a minimum the federal water quality standards. Unfortunately, the enforcement of these standards does not always guarantee the safety of our drinking water from chemical pollution. "Drinking water regulations are intended to reduce the risk of adverse health effects from exposure to contaminants," [1] not eliminate risk. Safety can only be assured by refocusing our environmental policies on *eliminating* the maximum concentration of pollutants in our drinking water instead of *limiting* selected pollutant concentrations to acceptable levels. To adopt such a policy will mean switching from passive enforcement of pollution based standards to actively removing pollutants predicated on technology-based water quality goals.

This approach is even more important today when terrorists threaten to poison our water supplies. Pollution of drinking water cannot be eliminated under the present federal approach nor can our government guarantee that our waters will be protected from chemical attack. Indeed, ". . . a more likely threat would probably come from contamination by a currently regulated chemical or microorganism or a common unregulated chemical or microorganism—contaminants that a water treatment system would not be designed to remove or inactivate because the agent would normally not be expected to occur" [2]. The delivery of the purest drinking water possible to the American public can only be achieved by adopting a national program to treat all drinking water to a level where pollutant concentrations are as close to "zero" as possible. Protecting the nation's public health has

129

always been this nation's highest priority. When the system fails to fully protect the public health, the system must be changed. However, the degree to which our nation's policy can be refocused to eliminate pollutants in our drinking water or for that matter ensure that all domestic water is safe from terrorist attacks will depend to a large extent on how much the public cares and how much the public is willing to pay.

Change will only come when the public demands a drinking water supply that contains the minimum concentrations of chemical, bacterial and radiological pollutants—in other words, truly pollution-free drinking water. Fortunately, this water can be achieved today with existing water treatment technologies. But as with so many environmental policies, the greatest barrier to the implementation of such a change is cost. Although cost can be a significant constraint, a drinking water policy based on best available treatment technologies can be implemented over time and within existing public agency budgets. The only question that remains is how long will the American public be willing to wait before these important changes are made?

POLLUTION-BASED POLICIES

If society is expected to accept drinking water that contains pollutants at "safe" levels[1], two critical criteria must be addressed and met. First of all, there should be a minimum of risk associated with the consumption of drinking water containing unregulated pollutants or pollutants at or below accepted standards. This assumption remains to be proven. Secondly, pollutant limitations that are established for our domestic water need to be strictly enforced by both state environmental agencies and the USEPA. The enforcement function, however, has major flaws. In August 2001, the USEPA's inspector general released a report on the enforcement effectiveness of the existing water pollution programs [3]. This report concluded that:

- At least 40 percent of the nation's waters do not meeting the standards states have set for them. Polluted runoff, both regulated and unregulated, is causing the majority of the nation's water quality problems.
- All 50 states found that their ability to identify water quality problems was constrained by a lack of non-point water quality data to determine the cause or source of water pollution. This lack of information has prevented states from setting non-point discharge standards. Because

[1]Which it has for the last 75 years.

no standards have been set, no definitive policy decisions have been made to control runoff from sources such as animal feed lots and agricultural lands.

- Environmental protection of water resources is principally enforced by state environmental agencies. The USEPA reports that when a state agency identifies a company that has violated a clean water regulation, they often fine the company too little and/or may never collect the fine. States frequently delay enforcement actions up to a year after a violation. Some states report that more than half of the facilities that violated a clean water regulation in 1999 continued to have the same violation in 2000.
- In order to fix the problem, the federal government needs to send more money to the states to enforce environmental programs. However, even with this financial support, some states are reluctant to enforce programs that may impact "small businesses and economically vital industries."

Because effective enforcement is one of the most important components of protecting the nation's drinking water, the deficiencies noted by the USEPA's inspector general casts serious doubts on the ability of state agencies to enforce federal water quality programs. Furthermore, the lack of monitoring data for a substantial number of unregulated chemicals found in both non-point and point sources of pollution is a significant flaw in the current enforcement program[2]. Without such data on the true distribution of unregulated chemicals in water resources, how can the USEPA determine if a water quality problem exists? This dilemma is a classic Catch-22 and is discussed in greater detail in Exhibit 4.1.

But even before more funds[3] are spent to enforce existing programs, an extensive and costly nationwide non-point water quality monitoring program must be implemented. Otherwise, enforcement cannot be as effective. If our basic water resources, do not Acome up to standard,@ then one must question the technical implementation of any program aimed at minimizing pollutants in our water supplies. Sadly, the ability to implement such a broad scale program is very close to being both economially and

---

[2]There is no comprehensive federal program to evaluate the occurrence and magnitude of unregulated pollutants in drinking water sources around the United States (i.e., only a very small number of compounds are currently monitored in the larger community water systems).

[3]Given the 2002 projected federal deficit, there is no reason to believe that any funds will be available for increased enforcement any time soon. This is particularly true when given the massive funding that is required for the "War on Terrorism."

Exhibit 4.1   Regulatory Policy and Monitoring for Unregulated Pollutants in Wastewater.

A city located in the Napa-Sonoma wine country of California recently proposed to divert a portion of its treated wastewater effluent for irrigation of local vineyards. Homeowners who obtained their drinking water from wells in the vicinity of the growers immediately sued the city in an attempt to block the implementation of the proposal. As environmental experts, we were retained by the homeowners to determine if the wastewater posed a pollution threat to the local groundwater.

Because the region is dominated by vineyards, the chemicals of concern in wastewater were pesticides specifically used for grapes in the Napa-Sonoma area (i.e., dimethoate, diphacinone, fenarimol, mancozeb, myclobutanil, oxyfluorfen and propargite). Of these pesticides, the city only monitored "occasionally" for diphacinone. In addition to the pesticides, the concern was raised that some unknown pharmaceuticals could pass through the wastewater treatment system and be discharged along with the wastewater effluent into the groundwater. Thus, the city was requested to monitor for (1) those pesticides that are specific to the Napa-Sonoma region and (2) those pollutants that may be an indicator that organic chemicals are passing through the city's wastewater treatment

technically impossible under the current system. This shortcoming is just one more reason why a technology-based water quality approach is so important.

Yet, there is another reason for a technology-based water quality program. With the advent of the Pentagon and World Trade Center attacks, there is now even a greater need to protect the public health from the terrorist attacks on our drinking water supplies. In order to addresses these pollution threats, existing water utility systems must be substantially improved. However, the ability to improve aging water treatment and delivery systems will be highly dependent upon both state and federal funding.

The only solution is to abandon our reliance upon drinking water standards as the means of protecting the public health and instead adopt a program that requires that drinking water be treated using the best available technologies to minimize the concentration of all pollutants.

(Exhibit 4.1, continued)

systems*. The selected indicator compounds included a wide range of phthalates, phenols and alkylbenzenesulfonates common to household products, caffeine, and selected pharmaceuticals (i.e., aspirin, ibuprofen, estrogen, clofibrate erythromycin and tetracycline).

If any of the indicator compounds were found in the effluent, then the treatment system might be allowing an unknown amount of various compounds to be discharged with the effluent. This finding would imply that the existing treatment system was not adequately designed to remove these pollutants. Thus, there would be no guarantee that the effluent was not a hazard to public health.

The city responded to the request by stating that there was no need to analyze for these chemicals in their wastewater because "criteria must exist to evaluate the toxicity of the chemicals." In other words, because no state or federal standards had been set for these chemicals (i.e., they are unregulated), there was no hazard nor need to monitor for them. This attitude clearly illustrates that after years of accepting "pollution-based standards," water quality agencies are still willing to allow toxic or hazardous chemicals in water simply because it is "politically" acceptable to pollute if the state or federal government have not set a standard.

*Since it would be impossible to analyze for all the potential pharmaceuticals that could pass through a wastewater treatment system, it was necessary to select a set of compounds that could be used as an indicator that the problem exists. These compounds were selected based on their occurrence in other municipal wastewater effluents (See Chapters 1 and 3).

A TECHNOLOGY-BASED POLICY

The least expensive option is to continue to place our trust in the current drinking water standards to protect human health (i.e., do no more than we are currently doing). As desirable as this alternative is, it would be wrong to assume that this 75-year-old policy will protect the public health just because it appears to have done so in the past. Times have changed and so should our reliance on drinking water standards.

The simplest alternative to protecting the public health from the threat of pollutants in drinking water would be to voluntarily install the best available technologies to minimize chemical pollution. This approach requires no change in governmental regulations since most if not all of the resulting contaminant levels would be below existing drinking water standards. Obviously, the beneficiaries should be responsible for the cost of these improvements. Those that can afford this level of treatment will also reap the benefit, while those unable to withstand the added financial burden will not.

To provide drinking water with the minimum concentration of chemical pollutants will require the federal government to change its current policy. This fact does not, however, prevent independent changes in policy by state governments in lieu of federal action since existing federal drinking water standards will not be violated.

Given the complexity and time (i.e., planning and engineering) required to address this problem, achieving a viable national policy based on minimizing pollutants based on technology will require (1) a philosophical change in thinking within the water industry, (2) a change in government policy and funding and (3) the need for consumers to provide the economic stimulus for higher quality treatment systems.

The Role of the Water Industry

First of all, it is important to realize that not all community water supplies and private sources of drinking water will need additional treatment to remove chemical pollutants. For example, water resources that originate in pristine watersheds will require little or no additional treatment to remove chemicals simply because the pollutant content is *de-minimis*. For example, the cities of San Francisco[4] and Seattle receive their water supply from protected watersheds. Similarly, deep groundwater resources that originate from pristine watersheds that are not recharged with impure wastewater or polluted rivers and streams or are not polluted by man's activities on the earth's surface (e.g., agriculture, leaking gasoline tanks, waste sites) may also require no additional treatment to remove chemical pollutants[5]. These types of areas need to be defined so that we focus on those water resources that will require additional treatment to remove chemical pollutants.

Water resources that are used as a source of drinking water and will require additional treatment to remove chemical include:

- Surface waters that pass through land masses that include mining, agricultural and urban development.

---

[4]However, the city of San Francisco needs to upgrade the pipeline from the watershed to the city at an estimated cost of over a four billion dollars just to maintain the integrity of this transmission system.

[5]This conclusion also assumes that there are not elevated concentrations of trace metals, such a arsenic, that can occur naturally in water resources.

- Surface water and groundwater that receive municipal or industrial wastewater discharges.
- Surface waters that are recharged by polluted groundwater.
- Groundwater under the influence of polluted surface waters (i.e., polluted river water that percolates into the groundwater).

Obviously, the level of pollution in each of the resources can vary widely depending upon the environmental conditions. Thus, water systems that obtain their water from any of these potentially polluted resources need to critically evaluate sources of chemical pollution that may contribute to the water resource and determine either how to control the sources of pollution or implement the necessary treatment technologies to guarantee their customers a source of drinking water that is as pollution free as technically possible.

The first step in providing drinking water with a minimum concentration of chemical pollutants is to assess the ability of water utilities to achieve this level of quality. One of the major hurdles facing utilities is the fact that only a very small fraction of the treated and distributed water is actually consumed as "drinking water." For example, approximately 90 percent of treated water is used by industry and commercial business, small agricultural businesses, landscape irrigation, firefighting, and a multitude of other uses outside the home (e.g., washing cars, walls and sidewalks, swimming pools, fountains, fish ponds, and lawn irrigation). These non-drinking water uses obviously do not have to be treated to current standards let alone to a zero level of pollution. Yet, a water utility has no choice because it distributes treated water without distinction as to use. As a result, water utilities who are already treating huge quantities of water far in excess of drinking water requirements would have to treat that same quantity to a much higher degree of purity.

An alternative would be the development of "dual use systems." In such an approach, a separate drinking water supply would be made available to home owners and businesses, while a lesser quality water system would be provided for agriculture, landscaping, firefighting and wastewater carriage. This approach would be a direct and much more cost effective method for obtaining high quality drinking water since a much smaller volume would require treatment using advanced treatment technologies. The drawback to such a system, however, would be in the cost of installing a dual distribution system. This separate system would require a massive expenditure of both federal and state funds to build new and parallel treatment plants and distribution systems. Clearly, privately owned water systems of which there are many could not afford such an undertaking without massive subsidies. Indeed, for the whole nation to adopt to this approach would be an impossible task in real time and within current budgets. However, this approach can be effectively planned and implemented in new communities, industrial parks and subdivisions. When these engineering and economic constraints are considered along with the need to address post-treatment pollution in the distribution system, many water utilities could not justify the production of a higher quality water to their consumers. Fortunately, there are other alternatives.

Professor Walter Weber of Michigan State University [4] has proposed the use of advanced water treatment technologies in a "satellite mode." These highly advanced treatment systems (i.e. coupling reverse osmosis with carbon filtration) would be used at the neighborhood level to improve the quality of the water coming from the central water treatment plant. The objective of using such a system would be to provide the highest quality water to a limited but specialized consumer base (i.e., housing subdivisions, apartment complexes or commercial districts). This approach has merit since a satellite system installed at a "point of need" would require a much shorter length of "parallel" distribution systems. According to Weber, the use of satellite systems is necessary since "potable water is of questionable quality . . . we're going to be facing the reality that water supply is, in fact, wastewater." Weber further points out that "although water treatment technologies continue to advance, they are too expensive to treat large quantities of water to potable[6] levels."

Another approach to reducing the volume of water requiring treatment to achieve minimum pollution water is to make even greater use of recycled water for those uses that do not require potable water. While this approach does not affect the quality of our drinking water, it does reduce the volume of water that must be treated to a near zero pollution level. In the western United States, recycled wastewater has been used for decades for watering lawns, gardens and golf courses[7]. In addition to these uses, a new trend towards the direct use of recycled wastewater for flushing toilets in commercial buildings has started. The Irvine Ranch Water District in Southern California began delivering recycled water to high-rise office buildings in Irvine's Jamboree Center complex in 1991 [5]. Recently the East Bay Municipal Utility District announced a similar use of recycled

---

[6]This comment is significant considering that Weber's statement does not even address water treatment to a zero pollution level.

[7]As discussed previously, the land application of wastewater always presents the possibility of groundwater pollution.

water in a 20-story office building in Oakland, California.[8] The building is fitted with a dual plumbing system to use treated wastewater for toilet flushing. Unlike the land application of wastewater, dual use systems in high-rise commercial buildings pose no threat to groundwater. Both of these approaches reduces the cost of producing high quality water because smaller volumes actually require treatment.

Ultimately, numerous alternatives exist for achieving a policy that will minimize the chemical pollutants in our drinking water. In some cases, water systems are already approaching or are at minimum chemical pollution in their product and all that needs to be done is to slightly increase the efficiency levels to further reduce pollutant levels. In other cases, major technical changes will have to be made in order to minimize the level of chemical pollutants. However, the basis of any policy must be predicated on the application of the best available technologies to attain a water product with minimum level of chemical pollutants.

The water industry, like most industries, is more often controlled and limited by economic rather than technical constraints. Concerns include meeting federal and state drinking water standards, delivering sufficient water to meet their costumer base, maintaining a satisfactory product quality level and staying within their budget. If these concerns are met, there is usually no incentive to provide an even higher quality product to its customers unless the community served is willing to pay the bill for this enhanced purity and added protection.

Given that the potential added cost to a water utility of implementing advanced treatment technologies should only add 15 to 25 percent to their budget, some communities could voluntarily make the necessary upgrade without state or federal support and past the added costs on to the consumer[9]. This situation is not an unlikely scenario. For example, as a result of the widespread pollution of drinking water by pesticides, some water utilities in the "corn belt" region of the United States have already upgraded water treatment facilities [6] at a significant cost to both the community and consumers.

With the implementation of the Stage 2 Disinfection Rule, those community water systems that exceed the new standards will have to reduce the amount of dissolved organic carbon in the raw water, switch to a non-chlorine/bromine disinfection systems (e.g., use ultraviolet light) or install advanced treatment technologies. In those cases where some degree of

---

[8]*San Francisco Chronicle*, Tuesday, July 31, 2001
[9]Based on communications with private and municipal water utility managers in California.

advanced treatment technologies are implemented to comply with the Stage 2 Rule, the cost of the additional improvements needed to reach minimum pollution in drinking water should be marginal. Thus, for some water utilities the leap to providing a truly pollution free drinking water to their customers is not that great. Ultimately, are water utilities capable of implementing the necessary technologies to attain drinking water that is close to pollution free as possible without state or federal funding support? The answer is yes, but are they willing to do it?

Another problem facing water utilities in their attempt to improve water quality through new treatment technologies is their aging distribution systems. Because of the threat of terrorist caused pollution, community water systems will also need funding to address security issues. The federal government is currently spending around $3 billion a year to repair our water supply infrastructure. The American Water Works Association reported in April 2001 [7] that an anticipated expenditure for drinking water infrastructure would cost approximately $151 billion over the next 20 years. Of this amount, approximately 25 percent of these funds would be used to upgrade treatment systems. To further confuse the cost projections, the American Water Works Association in July 2001 [8] quoted an USEPA report that projected expenditures of $250 billion over 30 years to upgrade Atens of thousands of miles of aging drinking water system pipes@ with little, or no, funds to address water treatment technology upgrades.

The American Water Works Association prepared an independent cost analysis that estimated a per capita expenditure of an additional $100 per customer[10] for each of the next 20 years. Based on this American Water Works Association assessment, approximately 68 percent of this estimated expenditure, or $102.6 billion, is needed immediately. Finally, the USEPA has estimated that $1 trillion will be required between 2010 and 2020 to meet current infrastructure objectives [9]. According to the 2000 Census, there are 105,480,101 households in the United States. Based on the USEPA estimate, the federal government would need to spend $9,480 per household over a ten year period to improve infrastructure. A more recent publication [10] projects a $50 billion annual expenditure to "build, operate and maintain needed drinking water facilities over the next 20 years." The report projects an average per capita share (assuming a U.S. population of 285 million people) of $175/person/year. These costs do not even guarantee the consumer that chemical pollutants will be absent "at the tap."

---

[10]Based on the AWWA model, this analysis only applies to consumers of the 20 large and medium size systems in the study.

These high expenditure projections are required just maintains the status quo. It should be emphasized that these cost projections have been developed and proposed by the water supply industry and the USEPA. They represent the combined wisdom of industry and government.

Given the difficulty water utilities face in (1) obtaining funds from federal and state governments and (2) raising the water bill rates of its costumers, it will be challenge to provide high quality drinking water. However, the water industry is capable of meeting this challenge. Because of the extensive investment in community water systems, it makes sense that these utilities should be the preferred and least costly distributor of drinking water. To facilitate this transformation, it is recommended that the water industry preform the following tasks:

- Educate local government and the communities they serve to the potential health risks and the solutions that can be adopted to meet community specific requirements. Through education, communities can make the necessary economic choices. By knowing the alternatives, many communities may be willing to pay a higher water bill to reap the potential health benefits.
- Communities can also allow for graduated water bills so that the lower income families will not bear undue cost increases.
- Begin the development of long term plans to rebuild water infrastructure into a dual use and/or satellite water supply system so that minimum pollution water can be delivered to customers at lower costs.
- Turning once again to the terrorist threat, it should be noted that it would be much more cost effective to design access security into new distribution pipelines than to attempt to retrofit existing distribution access. Such 'designed in' security would significantly reduce ability of terrorists to intentionally poison drinking water.
- A technology based water quality system will also require a new method of monitoring water treatment performance. Because many of the chemical pollutants that can occur in our water supplies cannot be realistically measured by only monitoring for specific chemicals or indicator chemicals, the water industry should be the leader in the development of fingerprinting for water resources both before and after treatment.
- Because maximum treatment will also remove natural minerals, the water industry should begin to evaluate methods of replacing minerals in treated drinking water (i.e., similar to the several bottled water products that mineralize their product after treatment).

In some instances, the water utility may not be able to provide service to an isolated group of individuals in the region served by the water utility. Under these circumstances, those homes or business might rely upon point-of-use water treatment systems. The water utility could offer to install (for a fee) a standardized point-of-use-system to those persons that would like to have such a system. Furthermore, for all those individuals that have point-of-use systems in the region of service, the water utility could provide both routine maintenance and monitoring services for a monthly or yearly fee to ensure that these systems are functioning properly.

The water industry should lead the way towards a technology based water quality program since the most feasible solutions are grounded in the communities that they serve. However, federal and state governments can also provide valuable assistance.

The Role of Government

It is not our wish that federal and state governments abandon the use of water quality criteria to protect the beneficial uses of water (i.e., aquatic environment, agriculture, industry and recreation) and to control the discharge of pollutants to the environment. It is critical, however, that federal and state governments abandon the concept of the drinking water standard as the means by which to protect the public health. The only true and complete protection is to have drinking water based on technology based standards of performance.

In order for the American public to have access to water that is as free of pollution as possible, the federal government must make a fundamental change in the approach by which "quality water" is provided to the consumer. Some of these changes will take more time than others. These should include (1) support to water utilities in their effort to provide higher quality water (2) assist in upgrading of system infrastructure, (3) support for small in-home and workplace water treatment systems, (4) require the bottled water and beverage industry to improve their guarantees of water purity, and (5) implement appropriate regulatory controls that are supportive of technology-based drinking water quality.

Support for Community Water Systems: The federal government needs to ensure that community water systems that serve transient populations (e.g., national parks, rest areas along interstate highways, federal lands, and reservations) are upgraded to provide pollution free drinking water. In some cases, the federal government may need to provide funding to community water systems that are unable to upgrade their facilities via rate increases. In addition to monetary support, the federal government should assist the water industry in developing chemical monitoring methods to fingerprint chemical pollution both before and after treatment to ensure that water treatment systems are functioning properly. Such procedures would drastically reduce monitoring costs as well as simplify regulatory oversight and related governmental costs.

Water Infrastructure: As discussed previously, the federal government needs to assist local communities with the planning and funding of dual and satellite water systems to replace old distribution systems as they are repaired and/or replaced. This involvement will obviously be long-term. As a result, the USEPA should be mandated to begin long-term program planning, fund research grants to universities to evaluate engineering alternatives, complete benefit-cost analysis, conduct evaluations of pollution free materials to be used in water systems, implement pilot scale projects for dual and satellite systems, develop and integrate in-situ methods of chemical monitoring for real-time treatment and control systems, and evaluate the effectiveness of treatment systems to remove specific chemicals and chemical compound classes.

Support for Small In-Home and Workplace Water Treatment Systems: Until the water infrastructure in the United States can be fully (or even partially) be replaced or community systems voluntarily upgrade their water treatment systems, the federal[11] government should sponsor programs that will promote the use of in-home and workplace water treatment systems. This step would be one of the quickest ways to help the American public attain drinking water that contains the minimum concentration of chemical pollution until such time that community based upgrades are installed. Such help is particularly important for that portion of the population that relies upon private sources of drinking water. Indeed, as of this writing, the federal government has already approved the use of point-of-use systems to meet the new arsenic standard.

An important part of this solution should be the provision of low-interest government loans and a tax credit program for the purchase of small water treatment systems. This approach has been practiced for many years with regard to homeowner efforts to reduce energy consumption. Tax cred-

---

[11]Although a federal program would be preferred, state or local governments could implement the same program.

it and rebate programs already exist for insulation, energy saving utilities and solar heating. Federal and state governments should also begin to provide small treatment systems for public buildings, facilities and federally sponsored low income housing. Installation of such systems in key federal buildings that might be the target of a terrorist attack is particularly important.

At a minimum, federal or state governments should offer support to those households and business that must rely on private sources of drinking water of questionable quality. Support should also be extended to those individuals that obtain their drinking water from community water systems that do not have a pristine source water and have not been upgraded to remove the maximum amount of chemical pollutants. The response to man-made chemicals should be no different than the government=s approach to point-of-use systems for arsenic removal.

Federal and state governments could also be more aggressive in safeguarding individuals that consume water currently not protected by best available treatment technologies by requiring that (1) all new home and building construction in these unprotected areas incorporate a water treatment system that removes the maximum concentration of chemical pollutants, (2) a minimum-pollution drinking water treatment system be installed whenever a private residence or building is remodeled (i.e., to the degree that a permit is required) and (3) a minimum pollution drinking water treatment system be installed whenever a private residence or building is sold prior to the sale. This type of program is consistent with other building code requirements that are usually implemented at the city or county level.

This approach is certainly not new and there are many examples where building codes have over time achieved the desired objectives. For example, this method has already been applied to smoke detectors, sprinkler systems, electrical wiring and plumbing, exterior wall insulation, seismic strengthening, flood plain zoning, and even the limitation in new construction of wood burning fireplaces. Providing minimum pollution drinking water through the mechanism of building codes will not guarantee a speedy implementation, but it will certainly move the nation more rapidly to a minimum pollution objective.

A program of this scope also requires the federal government to provide greater control of the testing, certification and availability of water purity data from standardized small treatment systems. This level of control is currently not available, although the National Sanitation Foundation does test and evaluate individual components of treatment systems. The USEPA should also fund university research on small system treatment design that will optimize chemical and biological water purity, while assuring easy maintenance and low operating costs.

State and federally sponsored programs that assist unprotected consumers in acquiring in-home/work place treatment systems will necessarily drive down the unit cost of these systems simply through mass production. Such capital and maintenance cost reduction will benefit all consumers and will further enable the public to install affordable water treatment technologies.

Bottled and Beverage Water: The federal government should require that all bottled water be properly labeled to reflect the environmental characteristics of the product, not the "real" and sometimes "invented" sources of bottled water. Current labels such as "artesian, spring, or mineral water" are environmentally meaningless. The following labels are recommended:

Pollutant-Free Water—Water treated to remove biologic organisms, insoluble and suspended matter, soluble inorganic salts and soluble organic compounds to as close to zero as possible using the best available technology. This product should be certified "pure" using chemical fingerprinting methods. No such product is currently available.

Treated Water—Water from community water systems or water from a community water system that has received additional treatment but not to the degree to be certified Pollutant-Free Water. A significant number of bottled water products are treated water. This includes products like "Aquafina," which comes from various community water systems around the United States and is subsequently treated using reverse osmosis.

Naturally Pure Water—Bottled water that originates from natural springs that occur in mountainous regions of the world with minimal human impact (i.e., wilderness designations or high alpine regions untouched by mining or towns), from deep aquifers whose water source originates from a wilderness area, or from deep thermal sources that have groundwater flow transport times in hundreds of years. This water is not guaranteed to be free of biologic organisms, man-made chemicals or toxic trace metals that could have been removed by the best available technology, but does represent the quality consistent with the source.

Potable Water—Natural sources of drinking water that do not meet the definition of "Naturally Pure Water" but meet current drinking water quality criteria with no guarantee that the product does not contain biologic organisms, man-made pollutants and toxic trace metals that could have been removed by the best available technology.

Currently, beverage water has no classifications that are similar to those used for bottled water. Therefore, beverage water should be pollutant free. In addition, no plastic bottles should be allowed for any bottled water or beverage unless the manufacturer of the product can (1) certify that no chemical component of the plastic bottle can leach into the water or beverage or (2) list the concentration[12] of each chemical compound that has been actually detected in the water or beverage as a result of being packaged in the products plastic container.

Proposed New Regulatory Programs: With the implementation of a technology-based water quality program, regulatory monitoring and notice programs can be greatly simplified since individual chemical components in drinking water will no longer be an issue. Monitoring compliance can rely on chemical finger printing methods to assess treatment performance. This approach will not only reduce the need for costly governmental oversight, but will also reduce monitoring costs for community water systems. As a result, regulatory agencies will need to modify compliance monitoring programs to augment and support technology based performance criteria.

A technology based treatment program is expected to generate a somewhat greater quantity of wastes resulting from the removal of greater amounts of chemicals. It should be both technically and economically feasible, for most communities, to dispose of these wastes back to the untreated water source[13]. This disposal method is especially true for those areas that meet existing beneficial use water quality standards for agriculture, industrial, recreation, and wildlife uses. The allowance of such discharges should to be part of any technology based drinking water quality program.

The Role of Concerned Citizens and Private-Sector Businesses

There are many potential solutions for attaining minimum pollution drinking water in our homes, in restaurants, in public places, in the workplace and in the bottled beverages we consume. Therefore, the most important role for those who appreciate the threat we all face is to (1) make your concerns known to your governmental representative, (2) join environmental

---

[12]The manufacturer must use the lowest detection level that can be achieved or use chemical finger printing methods to characterize the chemical impurities.

[13] In those areas where the removal of solids will not have a material influence on the beneficial use classification of the source water, the solids should be allowed to be discharged back into the source water.

organizations that promote solutions for reducing the pollution of our water resources and drinking water, (3) request that employers provide pollution free drinking water, and (4) for those individuals who do not have access to a minimum pollution water source and can afford the cost, install at least a point-of-use pollution free drinking water treatment system at the kitchen sink. By adopting these recommendations, the increased demand for small treatment systems should bring about a reduction in treatment system costs[14] that will make these systems more affordable and encourage their wider use. Finally, the installation of such small treatment systems will provide added protection against any intentional biologic or chemical attack on our water supplies.

The private sector can also make a significant contribution to improving the availability of purer water for their employees and customers by implementing a minimum pollution drinking water program without governmental mandate[15]. Considering just how many new housing developments, commercial facilities, shopping malls, recreation facilities, trailer parks, theaters and restaurants are planned and constructed each year in the United States, developers and businesses could make a major contribution to the improvement of drinking water quality by including the installation of minimum pollution systems and/or satellite treatment systems as an integral part of new construction projects. Not only can these costs be amortized over the life of the project, they will also help to reduce future system costs.

Concerned citizens can also begin the process of helping attain pollution free drinking water by political action and by installing in-home water treatment systems. To assist individual household or businesses in evaluating the appropriate components of minimum pollution treatment systems some of the basics of in-home treatment systems are described in Appendix O.

A BETTER POLICY

Today our drinking water is not certifiably free from chemical pollutants and is not protected from either further chemical pollution or possible terrorist acts. Protection cannot be assured under existing programs since they are founded on flawed policies and an inappropriate reliance upon outmoded and irrational drinking water standards. Key examples are listed in Exhibit 4.2. Water resources can be protected to a much greater degree if

---

[14]High-volume sales usually drive costs down as is the case with such organizations such as Costco, the retail/wholesale distributor.

[15]This response assumes that their source water does not already meet minimum pollution requirements.

Exhibit 4.2 Policy and Scientific Deficiencies

• Limits on specific chemicals in drinking water bear little or no relationship to a pollutant's effects on humans.
• Existing standards do not consider the health effects of multiple chemicals in the same water product.
• Knowledge of the animal toxicity of specific man-made chemicals cannot keep pace with the introduction of new chemical and biochemical compounds.
• No consistent list has been prepared of toxic chemicals that should be regulated in drinking water.
• Under the current system, it takes decades to regulate individual pollutants
• Unregulated chemicals found in drinking water are considered benign until proven otherwise.
• Risk assessment methods only estimate risk and cannot be validated for human populations.
• Regulatory programs that regulate pollution have caused the blurring of the boundary between wastewater and drinking water.

federal and state governments adopted the following key policies:
• Do not rely upon drinking water standards to protect public health.
• Require that drinking water that does not originate from pristine water resources, use the best available treatment technologies to remove chemical pollutants to the greatest degree possible.
• Implement programs that shift the primary responsibility of attaining maximum quality drinking water to community water systems with the cost of upgrading water treatment passed on to the consumer.
• When appropriate, provide economic assistance to community water systems to implement the best available water technologies.
• Implement programs that provide maximum quality drinking water to areas of the country that do not have access to quality sources of water. This goal could be accomplished by requiring that: (a) all new building construction include a point-of-entry treatment system for drinking water, (b) a point-of-entry water treatment system be installed whenever a private residence or building is remodeled (i.e., to the degree that a permit is required) and (c) a point-of-entry water treatment system be in-

stalled prior to the sale whenever a private residence or building is sold.
- Provide the impetus for reduction in the cost of point-of-use and point-of-entry systems by initiating a five year program to install these systems in federal facilities, buildings and federally sponsored housing projects that currently do not receive the highest quality drinking water.
- Implement long-term programs to assist local communities with planning and funding of dual and satellite water systems as an integral part of a program to repair and replace aging distribution systems. The dual use and satellite systems must be designed to provide pollution free water and be protected from terrorist incursion.
- Require that bottled water and beverages that use water as an additive be properly labeled to accurately reflect the environmental characteristics of their contents.

The degree to which our government agencies implement these policies, to a very large extent, will depend upon how much the public cares. Protection of drinking water from all forms of pollution must be a public goal given the potential hazards to the public health. For example,

- We know that today we are all consuming water containing both regulated and unregulated pollutants.
- We know that present government programs cannot guarantee pollution free water.
- We know that water supplies are not totally safe from terrorist actions.
- We know that the nation's entire water treatment and distribution system needs major overall, upgrading and replacement.
- We know that projections for system upgrade costs range upwards of a trillion dollars.
- We know that there will never be a comprehensive drinking water monitoring program that adequately addresses all unregulated pollutants.

Fortunately, for each one of these hazards there are defined and achievable solutions:

- We know that the technology exists to enable community water systems to produce and distribute a much higher quality water to their customers.
- We know that quality point-of-entry and point-of-use home water purification systems are available commercially.

- We know that dual-use systems have been implemented and have the capability of delivering high quality water at a lower cost.
- We know that the economics of furnishing high quality water to the majority of our population are within the boundaries of reality.
- We know that community drinking water systems can be designed to eliminate or significantly reduce the threat of terrorist acts.

Finally, we know that an enlightened and informed public can make high quality water a reality. All that remains is for us to demand that these solutions be considered, debated and reasonably implemented.

References

1. Pontius, Frederick W., "Regulatory Compliance Planning to Ensure Water Supply Safety," *Journal of the American Water Works Association*, p. 52, (March 2002)

2, Pontius, Frederick W., "Regulatory Compliance Planning to Ensure Water Supply Safety," *Journal of the American Water Works Association*, p.12, (March 2002).

3. USEPA, "State Enforcement of Clean Water Act Dischargers Can Be More Effective," Office of the Inspector General for Audit, Western Division, San Francisco, California, Report No. 2001-P-00013 (August 2001).

4. Landers, Jay, "Treatment Changes Needed to Ensure Sustainable Water Supply," *Water Environment & Technology* (December 2000).

5. Association of California Water Agencies—Water Facts, "Water Recycling," (Winter 2000 to 2001).

6. Cohen, Brian and Richard Wiles, "Tough to Swallow; How Pesticide Companies Profit from Poisoning America's Tap Water," *Environmental Working Group* (August 1997).

7. Scharfenaker, Mark, "Second National Need Survey Pegs Drinking Water Infrastructure costs at $150.9 Billion," *Journal of the American Water Works Association* (April 2001).

8. Scharfenaker, Mark, "Mythic monster aids first scientific assessment of nationwide water infrastructure needs," which discusses the AWWA report "Dawn of the Replacement Era: Reinvesting in Drinking Water Infrastructure," *Journal of the American Water Works Association* (July 2001).

9. Federal Water Review, "WIN Calls For Federal Infrastructure Assistance," *Association of Metropolitan Water Agencies* (March–April 2000).

10. Means, Edward G., et al., "The Coming Crisis: Water Institutions and Infrastructure," *Journal of the American Water Works Association* (January 2002).

Appendix A

Federal Drinking Water Standards

National Primary Drinking Water Standards for Chemicals[1]

Contaminant	Maximum Contaminant Level(MCL) (Micrograms per liter)[2]
Inorganic Compounds (19)	
Antimony	6.0
Arsenic	50.0
Barium	2000.0
Beryllium	4.0
Bromate	10.0
Cadmium	5.0
Chlorine gas (as Cl_2)	4000.0
Chlorine dioxide (as ClO_2)	800.0
Chlorite	1000.0
Chromium (total)	100.0
Copper	1300.0
Cyanide	200.0
Fluoride	4000.0
Lead	15.0
Mercury	2.0
Nitrate	10000.0
Nitrite	1000.0

[1]These are the legally enforceable standards that apply to public water systems (as of June, 2001).
[2]The maximum level of a contaminant in drinking water at which no known or anticipated adverse effect on the health effect of persons would occur, and which allows for an adequate margin of safety.

149

Contaminant	Maximum Contaminant Level(MCL) (Micrograms per liter)	
Selenium	50.0	
Thallium	2.0	

Organic Compounds (63)

Contaminant	MCL	
Acrylamide	1000.0	
Alachlor	2.0	Pesticide
Atrazine	3.0	Pesticide
Benzene	5.0	
Benzo(a)pyrene	0.2	
Carbofuran	40.0	Pesticide
Carbon tetrachloride	5.0	
Chloramines (as Cl_2)	4000.0	
Chlordane	2.0	Banned pesticide
Chlorobenzene	100.0	
Dichlorophenoxyacetic acid (2,4-D)	70.0	Pesticide
Dalapon	200.0	Pesticide
1,2-dibromo-3-chloropropane(DBCP)	0.2	Pesticide
O-Dichlorobenzene	600.0	
p-Dichlorobenzene	75.0	
1,2-Dichloroethane	5.0	
1,1-Dichloroethylene	7.0	
cis-1,2-Dichloroethylene	70.0	
trans-1,2-Dichloroethylene	100.0	
Dichloromethane	5.0	
1,2-Dichloropropane	5.0	
Di(2-ethylhexyl)adipate	400.0	
Di(2-ethylhexyl)phthalate	6.0	
Dinoseb	7.0	Pesticide
2,3,7,8-tetrachlorodibenzo-p-dioxin (TCDD)	0.00003	
Diaquat	20.0	Pesticide
Endothall	100.0	Pesticide
Endrin	2.0	Banned pesticide
Epichlorohydrin	20000.0	
Ethylbenzene	700.0	
Ethylene dibromide	0.05	
Glyphosate	700.0	Pesticide

Contaminant	Maximum Contaminant Level(MCL) (Micrograms per liter)	
Haloacetic acids (HAA5)[3]	60.0	
Heptachlor	0.4	Banned pesticide
Heptachlor epoxide	0.2	Banned pesticide
Hexachlorobenzene	1.0	
Hexachlorocyclopentadiene	50.0	
Lindane	0.2	Pesticide
Methoxychlor	40.0	Pesticide
Oxamyl(Vydate)	200.0	Pesticide
Polychlorinated biphenyls	0.5	
Pentachlorophenol	1.0	
Picloram	500.0	Pesticide
Simazine	4.0	Pesticide
Styrene	100.0	
Tetrachloroethylene	5.0	
Toluene	1000.0	
Total Trihalomethanes (TTHM)[4]	100.0	
Toxaphene	0.003	Pesticide
2,4,5-TP (Silvex)	50.0	Banned Pesticide
1,2,4-Trichlorobenzene	70.0	
1,1,1-Trichloroethane	200.0	
1,1,2-Trichloroethane	5.0	
Trichloroethylene	5.0	
Vinyl chloride	2.0	
Xylenes (total)	10000.0	

Asbestos fibers (MCL) 7 million fibers per liter (fibers must exceed 10 microns in length).

[3]The HAA5 compounds are dibromoacetic acid, dichloroacetic acid, monobromacetic acid, monochloroacetic acid, and trichloroacetic acid.

[4]The TTHM compounds are bromodichloromethane, bromoform, chloroform, and dibromochloromethane.

National Secondary Drinking Water Regulations

These standards are non-enforceable guidelines regulating contaminants that may cause cosmetic effects or aesthetic effects in drinking water. EPA recommends secondary standards to water systems but does not require systems to comply. However, individual states may choose to adopt them as enforceable standards.

Contaminant	Secondary Standard (Micrograms per liter)
Aluminum	0.05
Chloride	250
Copper	1.0
Fluoride	2.0
Iron	0.3
Manganese	0.05
Silver	0.1
Sulfate	250
Total Dissolved Solids	500
Zinc	5.0

Appendix B

USEPA 1998 Drinking Water Contaminant Candidate List

Acetochlor
Alachlor ESA (and acetanilide degradation products)
Aldrin
Aluminum
Boron
Bromobenzene
DCPA non-acid degradate
DCPA di-acid degradate
DDE
Diazinon
1,1-dichloroethane
1,1-dichloropropene
1,2-diphenylhydrazine
1,3-dichloropropane
1,3-dichloropropene
2,2-dichloropropane
2,4-dichlorophenol
Dieldrin
2,4-dinitrophenol
2,4-dinitrotoluene
2,6-dinitrotoluene
Disulfoton
Diuron
DPTC (s-ethyl-dipropylthiocarbanate)
Fonofos
Hexachlorobtadiene

p-isopropyltoluene (p-cymene)
Linuron
Manganese
Methyl bromide
Methyl-t-butyl ether (MTBE)
2-Methyl-phenol (o-cresol)
Metolachlor
Metrobuzin
Molinate
Naphthalene
Nitrobenzene
Organotins
Perchlorate
Prometon
RDX
Sodium
Sulfate
1,1,2,2-tetrachloroethane
Terbacil
Terbufos
Triazines (and degradation products)
2,4,6-trichlorophenol
1,2,4-trimethylbenzene
Vanadium

Appendix C

Suspected Endocrine-Disrupting Chemicals

Acetochlor (Contaminant Candidate List)*
Alachlor (Regulated)*
Aldicarb
Aldrin (Contaminant Candidate List)
Amitrole
Atrazine (Regulated)
Benomyl
Bifenthrin
Bromacil
Bromaxynil
Carbaryl
Carbofuran (Regulated)
Chlordane (Regulated)
Chlordecone
Chlorfentezine
Cyanazine
8-Cyhalothrin
2,4-D (Regulated)
DBCP (Regulated)
DCPA
DDT
DDE (Contaminant Candidate List)
DDD
Deltamethrin
Dicofol
Dieldrin (Contaminant Candidate List)
Dimethoate

Dinitrophenol (Contaminant Candidate List)
Dioxin (Regulated)
Endosulfan
Endrin
Ethiozin
Ethofenprox
Etridiazole
Fenarimol
Fenbuconazole
Fenitrothion
Fenvalerate
Fipronil
n-2-Fluorenylacetamide
Glufosinate-ammonium
alpha-HCH
beta-HCH
Heptachlor (Regulated)
Heptachlor epoxide (Regulated)
Hexachlorobenzene (Regulated)
Ioxynil
Lindane (gamama-BCH) (Regulated)
Linuron (Contaminant Candidate List)
Malathion
Mancozeb
Maneb
Methomyl
Methoxychlor (Regulated)
Metiram
Metribuzin
Mirex
Molinate (Contaminant Candidate List)
Nabam
Nitrofen
Oryzalin
Oxyacetmide/fluthamide
Oxychlordane
Paraquat
Parathion
Penachloronitrobenzene
Pendimethalin

Pentachlorophenol (Regulated)
Penta- to nonylphenols
Perchlorate (Contaminant Candidate List)
Phthalates
Photomirex
Picloram (Regulated)
Polychlorinated biphenyls (Regulated)
Prodiamine
Pronamide
Pyrethrins
Pyrethroids
Ronnel (fenchlorfos)
Simazine (Regulated)
Styrenes (Regulated)
2,4,5-T
Terbutryn
Thiazopyr
Toxaphene (Regulated)
Transnonachlor
Triadimefon
Tributyltin
Trichlorobenzene (Regulated)
Trifluralin
Vinclozolin
Zineb
Ziram

*Regulated chemicals are listed in Appendix A and the Contaminant Candidate List chemicals are in Appendix B.

Source: Pesticides News, No. 46 (December 1999) and Colborn, Theo, et al., *Our Stolen Future: Are We Threatening Our Fertility, Intelligence and Survival* (March 1997) and USEPA, "Perchlorate," Office of Water, Groundwater and Drinking Water (4/26/01).

Appendix D

U.S. Geological Survey Target Compounds

National Reconnaissance of Emerging Contaminants in U.S. Streams (2000)

Human Drugs

Prescription
Metformin (antidiabetic agent)
Cimetidine (antacid)
Raditidine (antacid)
Enalaprilat (antihypertensive)
Digoxin
Diltiazem (antihypertensive)
Fluoxetine (antidepressant)
Paroxetine (antidepressant)
Warfarin (anticoagulant)
Salbutamol (antiastmatic)
Gemfibrozil (antihyperlipidemic)
Dehydronifedipine (antianginal metabolite)
Digoxigenin (digoxin metabolite)

Non-Prescription
Acetaminophen (analgesic)
Ibuprofen (anti-inflammatory, analgesic)
Codeine (analgesic)
Caffeine (stimulant)
1,7-Dimethylxanthine (caffeine metabolite)
Cotinine (nicotine metabolite)

Veterinary and Human Antibiotics

Carbadox
Chlortetracycline
Ciprofloxacin
Doxycycline
Enrofloxacin
Erythromycin-H_2O (metabolite)
Lincomycin
Norfloxacin
Oxytetracycline
Roxithromycin
Sarafloxacin
Sulfachlorpyridazine
Sulfamerazine
Sulfamethazine
Sulfathiazole
Sulfadimethoxine
Sulfamethiazole
Sulfamethoxozole
Tetracycline
Trimethoprim
Tylosin
Virginiamycin

Sex and Steroidal Hormones

Biogenics
17b-Estradiol
17a-Estradiol
Estrone
Estriol
Testosterone
Progesterone
cis-Androsterone

Pharmaceuticals
17a-Ethynylestradiol (ovulation inhibitor)
Mestranol (ovulation inhibitor)
19-Norethisterone (ovulation inhibitor)

Equilenin (hormone replacement therapy)
Equilin (hormone replacement therapy)

Sterols
Cholesterol (fecal indicator)
3b-Coprostanol (carnivor fecal indicator)
Sigmastanol (plant sterol)

Industrial and Household Wastewater Products

Antioxidants
2,6-di-ter-Butylphenol
5-Methyl-1H-benzotriazole
Butylatehydroxyanisole (BHA)
Butylatedhydroxytoluene (BHT)
2,6-di-tert-Butyl-p-benzoquinone

Detergent metabolites
p-Nonylphenol
Nonylphenol monoethoxylate (NPEO1)
Nonylphenol diethoxylate (NPEO2)
Octylphenol monoethoxylate (OPEO1)
Octylphenol diethoxylate (OPEO2)

Fire retardants
Tri(2-chloroethyl)phosphate
Tri(dichlorisopropyl)phosphate

Insecticides
Diazion
Carbaryl
Chlorpyrifos
cis-Chlordane
N,N-diethyltoluamide (DEET)
Lindane
Methyl parathion
Dieldrin

Plasticizers
bis(2-Ethylhexyl)adipate
Ethanol-2-butoxy-phosphate
bis(2-Ethylhexyl)phthalate
Diethylphthalate
Triphenyl phosphate

Polycyclic aromatic hydrocarbons
Anthracene
Benzo(a)pyrene
Fluoranthene
Naphthalene
Phenanthrene
Pyrene

Others
Tetrachloroethylene (solvent)
Phenol (disinfectant)
1,4-Dichlorobenzene (fumigant)
Acetophenone (fragrance)
p-Cresol (wood preservative)
Phthalic andydride (used in plastic)
Bisphenol A (used in polymers)
Triclosan (antimicrobial disinfectant)

Appendix E

Regulated Pesticides in Food[1]

Pesticides with Residues Tolerances

Acephate
Acetochlor (1998 Drinking Water Contaminant Candidate List)
Alachlor (Regulated as a Primary Drinking Water Constituent)
Aldicarb
Allethrin (allyl homolog of cinerin I)
Aluminum tris (O-ethylphosphonate)
Ametryn
Aminoethoxyvinylglycine
4-Aminopyridine
Amitraz
Ammoniates for [ethylenebis-(dithiocarbamato] zinc and ethylenebis
    [dithiocarbamic acid] bimolecular and trimolecular cyclic anhydrosul-
    fides and disulfides
Arsanilic acid [(4-aminophenyl) arsonic acid
Asulam
Atrazine (Regulated as a Primary Drinking Water Constituent)
Avermectin B1 and it delta-8,9-isomer
Azoxystrobin
Barban
Benomyl
Benoxacor
Bentazon
Beta-([1,1-biphenyl]-4-yloxy)-alpha-(1,1-dimethylethyl)-1H-1,2,4-tria-
    zole-1-ethanol

---

[1]CFR Title 40, Part 180, Tolerances and Exemptions From Tolerances for Pesticide Chemicals in Food, 2000.

Beta-(4-Chlorophenoxy)-alpha-(1,1-dimethylethyl)-1H-1,2,4-triazole-1-
   ethanol
Bifenthrin
1,1-Bis(p-chlorophenyl)-2,2,2-trichloroethanol
Bromide (inorganic)
Bromacil
Bromoxynil
Buprofezin
Butylate
n-Butyl-N-ethyl-α· α· α-trifluoro-2,6-dinitro-p-toluidine
Cacodylic acid
Captafol
Captin
Carbaryl
Carbofuran (Regulated as a Primary Drinking Water Constituent)
Carbon disulfide
Carboxin
Carfentrazone-ethyl
Chlorfenapyr
Chlorimuron ethyl
2-[[4-chloro-6-(ethylamino)-s-triazin-2-yl]amino]-2-methylpropionitrile
Chloroneb
2-Chloro-N-isopropylacetanilide
p-Chlorophenoxyacetic acid
2-(m-Chlorophenoxy) propionic acid
Chloropyrifos
Chlorothalonil
Chlorpyrifos-methyl
Chlorsulfuron
CICP
Clethodim ((E)-2-[1-[[(3-chloro-2-propenyl)oxy]imino]propyl]-5-[2-
   (ethylthio)propyl]-3-hydroxy-2-cyclohexen-1-one
Clodinafop-propargyl
Clofencet
Clofentezine
Clomazone
Clopyralid
Cloquintocet-mexyl
Cloransulam-methyl
Copper

Copper carbonate
Coumaphos
Cyano(3-phenoxyphenyl)methyl-4-chloro-α-(1-methylethyl)
  benzeneacetate
Cyclanilide
Cyfluthrin
Cymoxanil
Cypermethrin and an isomer zeta-cypermethrin
Cyproconazole
Cyprodinil
Cyromazine
2,4-D (Regulated as a Primary Drinking Water Constituent)
Deltamethrin
Desmedipham
N,N-diallyl dichloroacetamide
Diazinon (1998 Drinking Water Contaminant Candidate List)
Dicamba
Dichlobenil
4-(Dichloroacetyl)-1-oxa-4-azaspiro[4,5]decane
4-(2,4-Dichlorophenoxy) butyric acid
1-[[2-(2,4-dichlorophenyl)-4-propyl-1,3-dioxolan-2-yl]methyl]-1H-1,2,4-
  triazole
2-(3,5-Dichlorophenyl)-2-(2,2,2-trichloroethyl)-oxirane
Dichlorvos
Diclofop-methyl
Diclosulam
O,O-Diethyl S-[2-(ethylthio)ethyl] phosphorodithioate
N,N-Diethyl-2-(4-methylbenzyloxy)ethylamine hydrochloride
N,N-Diethyl-2-(1-naphthalenyloxy)propionamide
Difenoconazole
Difenzoquat
Diflubenzuron
Diflufenzopyr
Dihydro-5-heptyl-2(3H)-furanone
2-[4,5-Dihydro-4-methyl-4-(1-methylethyl)-5-oxo-1H-imidazol-2-yl]-3-
  quinoline carboxylic acid
Dihydro-5-pentyl-2(3H)-furanone
S-(O,O-Diisopropyl phosphorodithioate) of N-(2-mercaptoethyl) ben-
  zenesulfonamide

Dimethenamid, 2-chloro-N-[(1-methyl-2methoxy)ethyl]-N-(2,4-dimethylthien-3-yl)-acetamide
Dimethipin
Dimethoate
Dimethomorph
2,2-Dimethyl-1,3-benzodioxol-4-ol methylcarbamate
O-[2-(1,1-Dimethylethyl)-5-pyrimidinyl]O-ethyl-O-(1-methylethyl) phosphorothioate
O,O-Dimethyl S-[(4-oxo-1,2,3-benzotriazin-3(4H)-yl)methyl]phosphorodithioate
O,O-Dimethyl S-[(4-oxo-1,2,3-benzotriazin-3(4H)-ylmethyl]phosphorodithioate
Dimethyl phosphate of 3-hydroxy-N-methyl-cis-crotonamide
Dimethyl phosphate of 3-hydroxy-N-N-dimethyl-cis-crotonamide
Dimethyl tetrachloroterephthalate
2,6-dimethyl-4tridecylmorpholine
4,6-Dinitro-o-cresol and its sodium salt
2,4-Dinitro-6-octylphenyl crotonate and 2,6-dinitro-4octylphenyl crotonate
Diphenamid
Diphenylamine
Dipropyl isocinchomeronate
Diquat
Diuron (1998 Drinking Water Contaminant Candidate List)
Dodine
Emamectin benzoate
Endosulfan
Endothall (Regulated as a Primary Drinking Water Constituent)
Esfenvalerate
Ethalfluralin
Ethephon
Ethion
Ethofumesate
Ethoprop
Ethoxyquin
5-Ethoxy-3-(trichloromethyl)-1,2,4-thiadiazole
O-Ethyl S-phenyl ethylphosphonodithioate
S-Ethyl cyclohexylethylthiocarbamate
S-Ethyl dipropylthiocarbamate
Ethylene oxide

S-Ethyl hexahydro-1H-azepine-1-carbothioate
S-[2-(Ethylsulfinyl)ethyl]O,O-dimethyl phospharothioate
Fenamiphos
Fenarimol
Fenbuconazole
Fenhexamid
Fenitrothion
Fenoxaprop-ethyl
Fenoxycarb
Fenpropathrin
Fenridazon
Fenthion
Ferbam
Fipronil
Fluazifop-butyl
Fludioxonil
Flumetsulam
Flumiclorac pentyl
Fluorine compounds
N-(4-fluorophenyl)-N-(1-methylethyl)-2-[[5-(trifluoromethyl)-1,3,4-
    thiadiazol-2-yl]oxy]
  Acetamide
Fluridone
Fluroxypyr 1-methylheptyl ester
Fluthiacet-methyl
Flutolanil(N-(3-(1-methylethoxy)phenyl)-2-(trifluoromethyl)benzamide)
Fluvalinate
Folpet
Formetanate hydrochloride
Furilazole
Glufosinate ammonium
Glyphosate (Regulated as a Primary Drinking Water Constituent)
Halosulfuron
Hexaconazole
Hexakis (2-methyl-2-phenylpropyl)distannoxane
Hexazinone
Hexythiazox
HOE-107892 (mefenpyr-diethyl)
Hydramethylnon
Hydrogen cyanide

Hydroprene
Imazalil
Imazamox
Imazapic-ammonium
Imazapyr
Imazethapyr, ammonium salt
Imidacloprid
Iprodione
Isoxaflutole
Kresoxim-methyl
Lactofen
Lambda-cyhalothrin
d-Limonene
Lindane (Regulated as a Primary Drinking Water Constituent)
Linuron (1998 Drinking Water Contaminant Candidate List)
Malathion
Maleic hydrazide
Mancozeb
Maneb
Mepiquat chloride
N-(Mercaptomethyl) phthalimide S-(O,O-dimethyl phosphorodithioate)
Metalaxyl
Metaldehyde
Methamidophos
Methanearsonic acid
Methidathion
Methomyl
Methoprene
Methoxychlor (Regulated as a Primary Drinking Water Constituent)
2-methyl-4chlorophenoxyacetic acid
4-(2-Methyl-4-chlorophenoxy) butyric acid
6-methyl-1,3-dithiolo [4,5-b] quinoxalin-2-one
Methyl 3-[(dimethoxyphosphinyl)oxy]butenoate
Methyl2-(4-isopropyl-4-methyl-5oxo-2-imidazolin-2yl)-p-toluate and
    methyl 6-(4-isopropyl-4-methyl-5-oxo-2-imidazolin-2-yl)-m-toluate
Metolachlor (1998 Drinking Water Contaminant Candidate List)
Metsulfuron methyl
Mineral oil
Myclobutanil
Naled

α-Naphthaleneacetamide
1-Naphthaleneacetic acid
N-1-Naphthyl phthalamicacid
Nicosulfuron, [3-pyridinecarboxamide, 2-((((4,6-dimethoxypyrimidin-2-yl)aminocarbonyl) Aminosulfonyl))-N,N-dimethyl]
Nicotine and Nicotine containing compounds
Nitrapyrin
Norflurazon
n-Octyl bicycloheptenedicarboximide
Orthoarsenic acid
Oryzalin
Oxadiazon
Oxadixyl
Oxamyl (Regulated as a Primary Drinking Water Constituent)
Oxyfluorfen
Oxytetracycline
Paraquat
Parathion
Pendimethalin
Pentachloronitrobenzene
Permethrin
Picloram (Regulated as a Primary Drinking Water Constituent)
Pirimiphos-methyl
Phenmedipham
o-Phenylphenol
Phorate
Phosphamidon
Phosphine
Phosphorothioic acid, 0,0-diethyl 0-(1,2,2,2-tetrachloroethyl) ester
Piperonyl butoxide
Prallethrin(RS)-2-methyl-4-oxo-3-(2-propynyl)cyclopent-2-enyl(IRS)-cis,trans-chrysanthemate
Primisulfuron-methyl
Procymidone
Profenofos
Prohexadione calcium
Propamocarb hydrochloride
Propanil
Propargite
Propazine

Propetamphos
S-Propyl butylethylthiocarbamate
S-Propyl dipropylthiocarbamate
Propylene oxide
Prosulfuron
Pymetrozine
Pyrazon
Pyrethrins
Pyridaben
Pyridate
Pyrimethanil
Pyriproxyfen
Pyrithiobac sodium
Quinclorac
Quizalofop ethyl
Resmethrin
Rimslfuron
Sethoxydim
Simazine (Regulated as a Primary Drinking Water Constituent)
Sodium dimethyldithiocarbamate
Sodium salt of acifluorfen
Sodium salt of fomesafen
Spinosad
Streptomycin
Sulfentrazone
Sulfosate (Sulfonium, trimethyl-salt with N-(phosphonomethyl)glycine
    (1:1))
Sulfosulfuron
Sulfur dioxide
Sulprofos
Synthetic isoparaffinic petroleum hydrocarbons
Tarta emetic
Tebuconazole
Tebufenozide
Tebuthiuron
Tefluthrin
Terbacil (1998 Drinking Water Contaminant Candidate List)
Terbufos (1998 Drinking Water Contaminant Candidate List)
Tetradifon
Tetrachlorvinphos

Tetraconazole
Thiabendazole
Thiazopyr
Thidiazuron
Thifensulfuron methyl (methy-3-[[[[(4-methoxy-6-methyl-1,3,5-triazin-2-
yl)amino]carbonyl] amino]sulfonyl]-2-thiophene carboxylate
Thiobencarb
2-(Thiocyano-methylthio)benzothiazole
Thiodicarb
Thiophanate-methyl
Tralkoxydim
Triadimefon
Thiram
Tralomethrin
Triasulfuron
Triazamate
Tribenuron
Tribuphos
Trichlorfon
S-2,3,3-Trichloroallyl diisopropylthiocarbamate
Triclopyr
Trifloxystrobin
Triflumizole
Trifluralin
Triflusulfuron methyl
Triforine
3,4,5-Trimethylphenyl methylcarbanate and 2,3,5-trimethylphenyl
methylcarbamate
Triphenyltin hydroxide
Vinclozolin
Zinc phosphide
Ziram

Appendix F

Dow Industrial Chemicals, Solvents and Dyes in 1938[1]

Acetanilid
Acetic anhydride
Acetylene tetrabromide
Aniline Oil
Anthranilic acid
Barium bromide
Benzoyl chloride
Bis phenol-A
Bromine
Bromoform
Cadmium bromide
Carbon bisulphide
Carbon tetrachloride
Caustic soda
Ciba blue
Ciba red
Ciba scarlet
Ciba violet
Chloracetyl chloride
Chloroform
Dichloracetic acid
Diethylaniline
Diethyl benzene
Diethylene glycol

Dimethylaniline
Diphenyl
Diphenyloxide
Dowanone blue
Dowanone yellow
Ethyl benzene
Ethyl bromide
Ethyl chloride
Ethylene chlorobromide
Ethylene dibromide
Ethylene dichloride
Ethylene glycol
Ethylene oxide
Ethyl monobromacetate
Ethyl monochloracetate
Ferric chloride
hexachlorethane
Hydrabromic Acid
Indigo
Isopropyl benzene
Magnesium bromide
Methyl monochloracetate
Midland vat blue
Monobrombenzene

[1]*Dow Industrial Chemicals and Dyes*, The Dow Chmeical Company, Midland, Michigan, 1938 and *Dow Industrial Solvents*, The Dow Chmeical Company, Midland, Michigan, 1938.

Monochloracetic acid
Monochlorobenzene
Muriatic acid
Orthodichlorbenzene
Orthachlor paranitraniline
Orthochlorphenol
Paradibrombenzene
Paradow
Paraphenetidin
Paraphenylphenol
Para tertiary butyl phenol
Perchlorethylene
Propylene dichloride
Phenol
Phenyl acetate
Phenyl glycine
Phenyl hydrazine
Phenyl methyl pyrazolone

Phthelimide
Potassium bromide
Salicylaldehyde
Salicylic acid
Sodium acetate
Sodium bromate
Sodium hydrosulphide
Sodium Sulphide
Sulphur chloride
Sulphur monochloride
1,1,2,2-Tetrachlorethane
Trichlorbezene
1,1,2-Trichlorethane
Trichlorethylene
Trimethylene bromide
Triphenyl phosphate
Zinc bromide

Appendix G

Total Toxic Organics
(Code of Federal Regulation, Title 40, Part 129)

Acenaphthene
Acenapthylene
Acrolein
Acrylonitrile
Aldrin
Anthracene
1,2-benzathracene (benzo(a)anthracene)
Benzene
Benzidine
Benzo(a)pyrene (3,4-benzopyrene)
3,4-Benzofluoranthene (benzo(b)fluoranthene)
11,12-benzofluoranthene (benzo(k)fluoranthene)
1,12-benzoperylene (benzo(ghi)perylene)
Bromoform (tribromomethane)
4-bromophenyl phenyl ether
Butyl benzyl phthalate
Di-n-butyl phthalate
Carbon tetrachloride (tetrachloromethane)
Chlordane
Chlorobenzene
Chlorodibromomethane
Chloroethane

Bis(2-chloroethoxy) methane
Bis(2-chloroethyl) ether
2-chloroethyl vinyl ether (mixed)
Chloroform (trichloromethane)
Bis(2-chloroisopropyl) ether
2-chloronaphthalene
2-chlorophenol
4-chlorophenyl phenyl ether
Chrysene
4,4-DDT
4,4-DDE (p,p-DDX)
4,4-DDD (p,p-TDE)
1,2,5,6-dibenzanthracene (dibenzo(a,h)anthracene)
1,2-dichlorobenzene
1,3-dichlorobenzene
1,4-dichlorobenzene
3,3-dichlorobenzidine
Dichlorobromomethane
1,2-dichloroethane
1,1-dichloroethane
1,1-dichloroethylene
1,2-trans dichloroethylene
2,4-dichlorophenol
1,2-dichloropropane
1,3-dichloropropylene (1,3-dichloropropene)
Dieldrin
Diethyl phthalate
2,4-dimethylphenol
Dimethyl phthalate
2,4-dinitrophenol
4,6-dinitro-o-cresol
2,4-dinitrotoluene
2,6-dinitrotoluene
1,2-diphenylhydrazine
Alpha-endosulfan

Beta-endosulfan
Endosulfan sulfate
Endrin
Endrin aldehyde
Ethylbenzene
Di (2-ethylhexyl) phthalate
Fluoranthene
Fluorene
Heptachlor
Heptachlor epoxide
Hexachlorobenzene
Hexachlorobutadiene
Alpha-hexachlorocyclohexane
Beta-hexachlorocyclohexane
Gamma-hexachlorocyclohexane
Delta-hexachlorocyclohexane
Hexachlorocyclopentadiene
Hexachloroethane
Isophorone
Methylene chloride (dichloromethane)
Methyl chloride (chloromethane)
Methyl bromide (bromomethane)
Naphthalene
Nitrobenzene
2-nitrophenol
4-nitrophenol
N-nitrosodimethylamine
N-nitrosodiphenylamine
N-nitrosodi-n-propylamine
Di-n-octyl phthalate
Parachlorometa cresol
Pentachlorophenol
Phenol
Phenanthrene
Polychlorinated biphenyls -1242 (Arochlor 1242)

Polychlorinated biphenyls -1254 (Arochlor 1254)
Polychlorinated biphenyls -1221 (Arochlor 1221)
Polychlorinated biphenyls -1232 (Arochlor 1232)
Polychlorinated biphenyls -1248 (Arochlor 1248)
Polychlorinated biphenyls -1260 (Arochlor 1260)
Polychlorinated biphenyls -1016 (Arochlor 1016)
Indeno (1,2,3-cd)pyrene (2,3-o-phenylene pyrene)
Pyrene
2,3,7,8-tetrachlorodibenzo-p-dioxin (TCDD)
1,1,2,2-tetrachloroethane
Tetrachloroethylene (perchloroethylene)
Toluene
Toxaphene
1,2,4-trichlorobenzene
1,1,1-trichloroethane
1,1,2-trichloroethane
Trichloroethylene
2,4,6-trichlorophenol
Vinyl chloride (chloroethylene)

Appendix H

Trace Organics in Lake Arrowhead, California, Wastewater

Benzaldehyde
1,2-Benzene dicarboxylic acid
Benzene, 1-methyl-4-2(methyl propyl)
Benzophenone
Benzothiazole, 2,2-(methylthio)
Bromodichloromethane
Butyl 2-methylpropyl ester
Cholestanol
3-Chloro-2-butanol
1,3,5-Cycloheptatriene
Cyclohexanone
Cyclohexanone, 4-(1,1-dimethyethyl)
Cyclopentanol 1,2-dimethyl-3-(methylethenyl)
Decahydro naphthalene
Decanal
Diacetate, 1,2-ethanediol
Dibromochloromethane
N,N-Diethyl-3-methyl benzamide
2,5-Dimethyl 3-hexanol
2,2-Dimethyl 3-pentanol
Ethanol, 2-butoxy-phosphate
Ethanone 1-(2-naphthalenyl)
Ethyl citrate
Fluoranthene*
Fluorene*

*Compounds on the Total Toxic Organics list (see Appendix D).

179

Heptanal
2-Heptanone,3 hydroxy-3 methyl
Hexadecanoic acid
Hexanal
3 Hexanol
3-Methoxy-3methyl-hexane
2-Methoxy-1-propanol
Nonanal
Octandecanoic acid
Octadiene,4,5 dimethyl-3,6 dimethyl
Octanal
Phenol 2,4 (bis(1,1-dimethylethyl))
Phenol 4,4(1,2-diethyl-1,2-ethanediyl)bisphenol nonyl
1-Phenyl ethanone
Propanic acid 2 methyl-2,2-dimethyl-1-(2hr...)
1-Propanol, 2-(2-hydroxypropoxy)
2-Propanone, 1-(1-cyclohexen-1-yl)
Tetradecanal
Tetradecanoic acid
Tribromomethane
Trichloromethane

Appendix I

Summary of Surface Water Data

The following data represent the frequency (as a percentage) that a specific pesticide was detected in a water sample.

| | | Basin Landuse | |
Compound	Agricultural	Urban	Mixed
Acetochlor			9.80
Acifluorfen	0.43		2.74
Alachlor	36.36	13.46	39.02
Aldicarb	0.32		
Aldicarb sulfoxide	0.11		
Atrazine	77.20	86.24	88.62
Atrazine, deethyl	52.80	50.46	62.20
Azinphos-methyl	2.61	0.93	1.22
Benfluralin	0.50	3.36	1.63
Bentazon	4.58	1.29	8.68
Bromacil	0.74	0.63	
Bromoxynil	0.64		
Butylate	7.70	1.83	8.16
Carbaryl	10.69	45.26	21.14
Carbofuran	11.99	2.75	9.35
Chlorothalonil			0.32
Chlorpyrifos	15.60	40.67	19.59
Cyanazine	27.67	8.26	45.93
2,4-D	11.62	13.50	9.13
Dacthal mono-acid	0.43		
2,4-DB	0.32	0.32	
DCPA	22.20	29.36	30.20

Compound	Basin Landuse		
	Agricultural	*Urban*	*Mixed*
DDE	6.30	1.22	4.90
Diazinon	16.90	74.85	45.31
Dicamba	0.53	0.32	1.83
Dichlobenil	0.11	1.59	0.90
Dichlorprop	0.43	0.96	0.46
Dieldrin	6.90	3.67	3.27
2,4-Diethylaniline	4.50	0.61	4.90
Dinoseb	0.11		
Disulfoton	0.50	0.92	
Diuron	7.96	13.02	9.46
DNOC		0.32	
EPTC	25.13	3.98	22.86
Ethalfluralin	3.30		0.41
Ethiprop	3.40	1.83	3.25
Fenuron	0.11		0.45
Fluometuron	2.86		
Fonofos	5.80	2.14	10.20
HCH, alpha	0.40		0.41
HCH, gamma	1.90	0.92	2.86
Linuron	3.40	1.53	2.04
Malathion	5.60	19.57	8.16
MCPA	1.81	4.82	
Methiocarb	0.11		0.45
Methomyl	0.96	0.32	
Methyl parathion	0.90	0.31	0.41
Metolachlor	73.23	58.59	81.30
Metribuzin	13.70	6.73	14.29
Molinate	4.90	0.92	2.04
Napropamide	7.59	1.83	3.66
Neburon		0.32	
Norflurazon	0.64	0.32	
Oryzalin	0.74	3.81	0.45
Parathion	0.20	0.31	
Pebulate	2.10		4.08
Pendimethalin	11.00	19.57	8.57
Permethrin, cis	0.30		0.82
Phorate	0.10		0.41
Prometon	34.97	83.79	61.79

Compound	Basin Landuse		
	Agricultural	Urban	Mixed
Pronamide	2.20	0.31	2.85
Propachlor	3.00	1.83	3.66
Propanil	1.90	2.75	
Propargite	5.30	0.31	3.27
Propham	0.11	0.63	
Propoxur	0.22	0.33	
Simazine	61.74	87.77	82.93
Tebuthiuron	15.88	31.19	34.15
Terbacil	4.64	1.23	3.28
Terbufos	0.20		0.41
Thiobencarb	3.10	1.22	1.22
Triallate	8.50	0.31	6.94
Triclopyr	1.17	2.25	
Trifluralin	17.50	6.42	15.92

Appendix J

Summary of Shallow Groundwater Data

The following data represent the frequency (as a percentage) that a specific pesticide was detected in a water sample.

	Basin Landuse	
Compound	Agricultural	Urban
Acetochlor	0.25	
Alachlor	3.14	0.33
Aldicarb sulfoxide	0.34	
Atrazine	43.72	24.25
Atrazine, deethyl	42.27	20.27
Azinphos-methyl	0.44	
Benfluralin	0.11	0.33
Bentazon	1.01	0.35
Bromacil	1.23	2.42
Butylate	0.32	
Carbaryl	0.32	1.33
Carbofuran	0.76	0.66
Chlorpyrifos	0.54	
Cyanazine	1.73	1.33
2,4-D	0.45	1.04
2,4-DB	0.11	
Dacthal mono-acid	0.11	
DBCP	1.35	
DCPA	1.41	
DDE	3.68	1.99
Diazinon	0.54	1.66
Dicamba	0.11	0.35
Dichlobenil	0.33	0.35

| Compound | Basin Landuse | |
	Agricultural	Urban
1,2-dichloropropane	1.75	0.33
Dichlorprop	0.11	
Dieldrin	0.97	5.65
2,4-Diethylaniline	1.19	
Dinoseb	0.34	
Diuron	2.34	2.77
EDB	0.41	
EPTC	1.62	
Ethalfluralin	0.33	
Ethiprop	0.11	
Fluometuron	0.45	
HCH, alpha	0.11	
Linuron	0.11	
Malathion	0.32	
Methyl parathion	0.22	
Metolachlor	18.37	9.97
Metribuzin	3.46	1.66
Norflurazon	0.11	
Oryzalin		0.69
Pebulate	0.32	
Pendimethalin	0.22	
Permethrin, cis	0.22	
Picloram	0.11	
Prometon	13.40	27.24
Pronamide	0.22	
Propachlor	0.32	
Propanil	0.76	0.33
Propargite	0.11	
Propoxur		0.35
Simazine	22.38	15.28
Tebuthiuron	1.73	6.31
Terbacil	0.87	0.34
Terbufos	0.11	0.33
Triallate	0.32	0.33
1,2,3-trichloropropane	1.08	
Trifluralin	0.43	0.66

Appendix K

Unregulated Pollutants Discharged to or Identified in Water Resources

Acetone
Acetochlor
Acifluorfen
Alcohol ethoxylate
Aldicarb
Aldicarb sulfoxide
Alkylbenzenesulfonates
Alkyl diamine
Alkyl lead
Aluminum
Atrazine, deethyl
Azinphos-methyl
Benfluralin
Bentazon
Benzaldehyde
1,2-Benzene dicarboxylic acid
Benzene, 1-methyl-4-2(methyl propyl)
Benzoic acid
Benzophenone
Benzothiazole, 2,2-(methylthio)
Benzyl alcohol
Bromacil
Bromoxynil
Butyl methyl ketone
Butyl 2-methylpropyl ester
Butylate
Caffeine
Carbamazepine

Carbofuran
Cholestanol
4-Chloroaniline
3-Chloro-2-butanol
p-Chloro-m-cresol
2-Chloroethyl vinyl ether
4-Chlorophenol
Chlorothalonil
Chlorpropamide
Chlorpyrifos
Chromium, hexavalent
Cyanazine
1,3,5-Cycloheptatriene
Cyclohexanone
Cyclohexanone, 4-(1,1-dimethyethyl)
Cyclopentanol 1,2-dimethyl-3-(methylethenyl)
Dacthal mono-acid
2,4-DB
Decahydro naphthalene
Decanal
Diacetate, 1,2-ethanediol
Diazinon
Dibenzofuran
Dicamba
Dichlobenil
1,2-Dichloropentane
Dichlorprop
Dicyclohexylamine
2,4-Diethylaniline
N,N-Diethyl-3-methyl benzamide
2,5-Dimethyl 3-hexanol
2,2-Dimethyl 3-pentanol
Disulfoton
Diuron
DNOC
EPTC
Ethalfluralin
Ethanol, 2-butoxy-phosphate
Ethanone 1-(2-naphthalenyl)
Ethiprop
Ethyl citrate
Ethylenediaminetetraacetic acid (EDTA)
Fenuron
Fluometuron

Fonofos
HCH, alpha
HCH, gamma
Heptanal
2-Heptanone,3 hydroxy-3 methyl
Hexadecanoic acid
Hexanal
3 Hexanol
Isobutyl methyl ketone
Linuron
Malathion
MCPA
Meprobomate
Methiocarb
Methomyl
3-Methoxy-3methyl-hexane
2-Methoxy-1-propanol
Methylene-blue
Methyl ethyl ketone
2-Methylnaphthalene
Methyl parathion
2-Methylphenol
4-Methylphenol
Methyl tertiary-butyl ether (MTBE)
Metolachlor
Metribuzin
Molinate
Mirex
Neburon
Nickel
2-Nitroaniline
3-Nitroaniline
4-Nitroaniline
Nonanal
Nonylphenol
Norflurazon
Octachlorostyrene
Octadiene,4,5 dimethyl-3,6 dimethyl
Octandecanoic acid
Octanal
Oryzalin
Parathion
Pebulate
Pendimethalin

Pentobarbital
Perfluoro-octanyl sulfonate (PFOS)
cis-Permethrin
Phenol
Phenol 2,4 (bis(1,1-dimethylethyl))
Phenol 4,4(1,2-diethyl-1,2-ethanediyl)bisphenol nonyl
Phensuximide
1-Phenyl ethanone
Phorate
Piperonyl butoxide
Polyethylene glycol
Propanic acid 2 methyl-2,2-dimethyl-1-(2hr...)
1-Propanol, 2-(2-hydroxypropoxy)
2-Propanone, 1-(1-cyclohexen-1-yl)
Propargite
Propham
Propoxur
Sodium dodecylbenzenesulfonate
Sulfonamide
Tebuthiuron
Terbacil
Terbufos
Tetradecanal
Tetradecanoic acid
Thallium
Thiobencarb
m-toluate
Triallate
Trichlopyr
Trichlorofluoromethane
2,4,5-Trichlorophenol
Trihaloalkylphosphates
Trifluralin
Trimethyltriazinetrione
Vinyl acetate

Appendix L

Chemicals Known to the State of California to Cause Cancer or Reproductive Toxicity

(As of April 20, 2001)

Chemicals Known to Cause Cancer

A-alpha-C (2-Amino-9H-pyrido[2,3-b]indole)
Acetaldehyde
Acetamide
Acetochlor
2-Acetylaminofluorene
Acifluorfen
Acrylamide
Acrylonitrile
Actinomycin D
Adriamycin (Doxorubicin hydrochloride)
AF-2;[2-(2-furyl)-3-(5-nitro-2-furyl)]acrylamide
Aflatoxins
Alachlor
Aldrin
2-Aminoanthraquinone
p-Aminoazobenzene
ortho-Aminoazotoluene
4-Aminobiphenyl (4-aminodiphenyl)
1-Amino-2,4-dibromoanthraquinone
3-Amino-9-ethylcarbazole hydrochloride
2-Aminofluorene
1-Amino-2-methylanthraquinone
2-Amino-5-(5-nitro-2-furyi)-1,3,4-thiadiazole
4-Amino-2-nitrophenol
Amitrole

Analgesic mixtures containing phenacetin
Aniline
Aniline hydrochloride
ortho-Anisidine
ortho-Anisidine hydrochloride
Antimony oxide (Antimony trioxide)
Aramite
Arsenic (inorganic arsenic compounds)
Asbestos
Auramine
Azacitidine
Azaserine
Azathioprine
Azobenzene
Benz[a]anthracene
Benzene
Benzidine [and its salts]
Benzidine-based dyes
Benzo[b]fluoranthene
Benzoo]fluoranthene
Benzo[k]fluoranthene
Benzofuran
Benzo[a]pyrene
Benzotrichloride
Benzyl chloride
Benzyl violet 4B
Beryllium and beryllium compounds
2,2-Bis(bromomethyl)-1,3-propanediol
Bis(2-chloroethyl)ether
N,N-Bis(2-chloroethyl)-2-naphthylamine (Chlornapazine)
Bischloroethyll nitrosourea (BCNU) (Carmustine)
Bis(chloromethyl)ether
Bis(2-chloro-l-methylethyl)ether, technical grade
Bitumens
Bracken fern
Bromodichloromethane
Bromoethane
Bromoform
1,3-Butadiene
IIA-Butanediol dimethanesulfonate (Busulfan)
Butylated hydroxyanisole
beta-Butyrolactone
Cacodylic acid
Cadmium and cadmium compounds

Caffeic acid
Captafol
Captan
Carbazole
Carbon tetrachloride
Carbon-black extracts
Ceramic fibers (airborne particles of respirable size)
Certain combined chemotherapy for lymphomas
Chlorambucil
Chloramphenicol
Chlordane
Chlordecone (Kepone)
Chlordimeform
Chlorendic acid
Chlorinated paraffins (Average chain length, C12; approximately 60 percent chlorine by
  weight)
p-Chloroaniline
p-Chloroaniline hydrochloride
Ghlered-AFememethaRe Delisted October 29, 1999
Chloroethane (Ethyl chloride)
1-(2-Chloroethyl)-3-cyclohexyl-l-nitrosourea (CCNU) (Lomustine)
1-(2-Chloroethyl)-3-(4-methylcyclohexyl)-1 -nitrosourea (Methyl-CCNU)
Chloroform
Chloromethyl methyl ether (technical grade)
3-Chloro-2-methylpropene
1 -Chloro-4-nitrobenzene
4-Chloro-ortho-phenylenediamine
p-Chloro-o-toluidine
p-Chloro-o-toluidine, strong acid salts of
5-Chloro-o-toluidine and its strong acid salts
Chloroprene
Chlorothalonil
Chlorotrianisene
Chlorozotocin
Chromium (hexavalent compounds)
Chrysene
C.I. Acid Red 114
C.I. Basic Red 9 monohydrochloride
C.I. Direct Blue 15
C.I. Direct Blue 218
C.I. Solvent Yellow 14
Ciclosporin (Cyclosporin A; Cyclosporine)
Cidofovir
Cinnamyl anthranilate

Cisplatin
Citrus Red No. 2
Clofibrate
Cobalt metal powder
Cobalt [11] oxide
Cobalt sulfate heptahydrate
Conjugated estrogens
Creosotes
para-Cresidine
Cupferron
Cycasin
Cyclophosphamide (anhydrous)
Cyclophosphamide (hydrated)
Cytembena
D&C Orange No. 17
D&C Red No. 8
D&C Red No. 9
D&C Red No. 19
Dacarbazine
Daminozide
Dantron (Chrysazin; 1,8-Dihydroxyanthraquinone)
Daunomycin
DDD (Dichlorodiphenyidichloroethane)
DDE (Dichlorodiphenyidichloroethylene)
DDT (Dichlorodiphenyltrichloroethane)
DDVP (Dichlorvos)
N,N'-Diacetylbenzidine
2,4-Diaminoanisole
2,4-Diaminoanisole sulfate
4,4'-Diaminodiphenyl ether (4,4'-Oxydianiline)
2,4-Diaminotoluene
Diaminotoluene (mixed)
Dibenz[a,h]acridine
Dibenz[a,j]acridine
Dibenz[a,h]anthracene
7H-Dibenzo[c,g]carbazole
Dibenzo[a,e]pyrene
Dibenzo[a,h]pyrene
Dibenzo[a,i]pyrene
Dibenzo[a,l]pyrene
1,2-Dibromo-3-chloropropane (DBCP)
2,3-Dibromo-1 -propanol
Dichloroacetic acid
p-Dichlorobenzene

3,3'-Dichlorobenzidine
3,3'-Dichlorobenzidine dihydrochloride
1,4-Dichloro-2-butene
3,3'-Dichloro-4,4'-diaminodiphenyl ether
M-Dichloroethane
Dichloromethane (Methylene chloride)
1,2-Dichloropropane
1,3-Dichloropropene
Dieldrin
Dienestrol
Diepoxybutane
Diesel engine exhaust
Di(2-ethylhexyl)phthalate
1,2-Diethylhydrazine
Diethyl sulfate
Diethylstilbestrol (DES)
Diglycidyl resorcinol ether (DGRE)
Dihydrosafrole
Diisopropyl sulfate
3,3'-Dimethoxybenzidine (ortho-Dianisidine)
3,3'-Dimethoxybenzidine dihydrochloride (ortho-Dianisidine dihydrochloride)
Dimethyl sulfate
4-Dimethylaminoazobenzene
trans-2-[(Dimethylamino)methylimino]-5-[2-(5-nitro-2-furyl)vinyl]-1,3,4-oxadiazole
7,12-Dimethylbenz(a)anthracene
3,3'-Dimethylbenzidine (ortho-Tolidine)
3,3'-Dimethylbenzidine dihydrochloride
Dimethylcarbamoyl chloride
1,1-Dimethylhydrazine (UDMH)
1,2-Dimethylhydrazine
Dimethylvinylchloride
3,7-Dinitrofluoranthene
3,9-Dinitrofluoranthene
1,6-Dinitropyrene
1,8-Dinitropyrene
Dinitrotoluene mixture, 2,442,6
2,4-Dinitrotoluene
2,6-Dinitrotoluene
Di-n-propyl isocinchomeronate (MGK Repellent 326)
1,4-Dioxane
Diphenylhydantoin (Phenytoin)
Diphenylhydantoin (Phenytoin), sodium salt
Direct Black 38 (technical grade)
Direct Blue 6 (technical grade)

Direct Brown 95 (technical grade)
Disperse Blue 1
Epichlorohydrin
Erionite
Estradiol 17B
Estragole
Estrone
Estropipate
Ethinylestradiol
Ethoprop
Ethyl acrylate
Ethyl methanesulfonate
Ethyl-4,4'-dichlorobenzilate
Ethylene dibromide
Ethylene dichloride (1,2-Dichloroethane)
Ethylene oxide
Ethylene thiourea
Ethyleneimine
Fenoxycarb
Folpet
Formaldehyde (gas)
2-(2-Formy[hydrazino)-4-(5-nitro-2-furyl)thiazole
Furan
Furazolidone
Furmecyclox
Fusarin C
Ganciclovir sodium
Gasoline engine exhaust (condensates/extracts)
Gernfibrozil
Glasswool fibers (airborne particles of respirable size)
Glu-P-1 (2-Amino-6-methyldipyrido[1,2 a:3',2'-d]imidazole)
Glu-P-2 (2-Aminodipyrido[1,2-a:3',2'-d]imidazole)
Glycidaldehyde
Glycidoll
Griseofulvin
Gyromitrin (Acetaldehyde methylformylhydrazone)
HC Blue 1
Heptachlor
Heptachlor epoxide
Hexachlorobenzene
Hexachlorocyclohexane (technical grade)
Hexachlorodibenzodioxin
Hexachloroethane
Hexamethylphosphoramide

Hydrazine
Hydrazine sulfate
Hydrazobenzene (1,2-Diphenylhydrazine)
Indeno [1,2,3-ccQpyrene
Indium phosphide
IQ (2-Amino-3-methylimidazo[4,5-flquinoline)
Iprodione
Iron dextran complex
Isobutyl nitrite
Isoprene
Isosafrole
Isoxaflutole
Lactofen
Lasiocarpine
Lead acetate
Lead and lead compounds
Lead phosphate
Lead subacetate
Lindane and other hexachlorocyclohexane isomers
Lynestrenol
Mancozeb
Maneb
Me-A-alpha-C (2-Amino-3-methyl-9H-pyrido[2,3-b]indoie)
Medroxyprogesterone acetate
MeIQ(2-Amino-3,4-dimethylimidazo[4,5-flquinoline)
MeIQx(2-Amino-3,8-dimethylimidazo[4,5-flquinoxaline)
Melphalan
Merphalan
Mestranol
Metham sodium
8-Methoxypsoralen with ultraviolet A therapy
5-Methoxypsoralen with ultraviolet A therapy
2-Methylaziridine (Propyleneimine)
Methylazoxymethanol
Methylazoxymethanol acetate
Methyl carbarnate
3-Methylcholanthrene
5-Methylchrysene
4,4'-Methylene bis(2-chloroaniline)
4,4'-Methylene bis(N,N-dimethyl)benzenamine
4,4'-Methylene bis(2-methylaniline)
4,4'-Methylenedianiline
4,4'-Methylenedianiline dihydrochloride
Methylhydrazine and its salts

Methyl iodide
Methylmercury compounds
Methyl methanesulfonate
2-Methyl-l-nitroanthraquinone (of uncertain purity)
N-Methyl-N'-nitro-N-nitrosoguanidine
N-Methylolacrylamide
Methylthiouracil
Metiram
Metronidazole
Michler's ketone
Mirex
Mitomycin C
Monocrotaline
5-(Morpholinomethyl)-3-[(5-nitro-furfurylidene)- amino]-2-oxazolidinone
Mustard Gas
MX (3-chloro-4-dichloromethyl-5-hydroxy-2(5H)-furanone)
Nafenopin
Nalidixic acid
1-Naphthylamine
2-Naphthylamine
Nickel and certain nickel compounds
Nickel carbonyl
Nickel refinery dust from the pyrometallurgical process
Nickel subsulfide
Niridazole
Nitrilotriacetic acid
Nitrilotriacetic acid, trisodium salt monohydrate
5-Nitroacenaphthene
5-Nitro-o-anisidine
o-Nitroanisole
Nitrobenzene
4-Nitrobiphenyl
6-Nitrochrysene
Nitrofen (technical grade)
2-Nitrofluorene
Nitrofurazone
1-[(5-Nitrofurfurylidene)-amino]-2-imidazolidinone
N-[4-(5-Nitro-2-furyl)-2-thiazolyl]acetamide
Nitrogen mustard (Mechlorethamine)
Nitrogen mustard hydrochloride (Mechlorethamine hydrochloride)
Nitrogen mustard N-oxide
Nitrogen mustard N-oxide hydrochloride
Nitromethane
2-Nitropropane

1-Nitropyrene
4-Nitropyrene
N-Nitrosodi-n-butylamine
N-Nitrosodiethanolamine
N-Nitrosodiethylamine
N-Nitrosodimethylamine
p-Nitrosodiphenylamine
N-Nitrosodiphenylamine
N-Nitrosodi-n-propylamine
N-Nitroso-N-ethylurea
3-(N-Nitrosomethylamino)propionitrile
4-(N-Nitrosomethylamino)-l -(3-pyridyl)l -butanone
N-Nitrosomethylethylamine
N-Nitroso-N-methylurea
N-Nitroso-N-methylurethane
N-Nitrosomethylvinylamine
N-Nitrosomorpholine
N-Nitrosonornicotine
N-Nitrosopiperidine
N-Nitrosopyrrolidine
N-Nitrososarcosine
o-Nitrotoluene
Norethisterone (Norethindrone)
Norethynodrel
Ochratoxin A
Oil Orange SS
Oral contraceptives, combined
Oral contraceptives, sequential
Oxadiazon
Oxazepam
Oxymetholone
Oxythioquinox
Panfuran S
Pentachlorophenol
Phenacetin
Phenazopyridine
Phenazopyridine hydrochloride
Phenesterin
Phenobarbital
Phenolphthalein
Phenoxybenzamine
Phenoxybenzamine hydrochloride
o-Phenylenediamine and its salts
Phenyl glycidyl ether

Phenylhydrazine and its salts
o-Phenylphenate, sodium
o-Phenylphenol
PhiP(2-Amino-1 -methyl-6-phenylimidazol[4,5-b]pyddine)
Polybrominated biphenyls
Polychlorinated biphenyls
Polychlorinated biphenyls (containing 60 or more percent chlorine by molecular weight)
Polychlorinated dibenzo-p-dioxins
Polychlorinated dibenzofurans
Polygeenan
Ponceau MX
Ponceau 3R
Potassium bromate
Primidone
Procarbazine
Procarbazine hydrochloride
Procymidone
Progesterone
Pronamide
Propachlor
1,3-Propane sultone
Propargite
beta-Propiolactone
Propylene oxide
Propylthiouracil
Quinoline and its strong acid salts
Radionuclides
Reserpine
Residual (heavy) fuel oils
Safrole
Salicylazosulfapyridine
Selenium sulfide
Soots, tars, and mineral oils
Spironolactone
Stanozolol
Sterigmatocystin
Streptozotocin (streptozocin)
Styrene oxide
Sulfallate
Tamoxifen and its salts
Terrazole
Testosterone and its esters
2,3,7,8-Tetrachlorodibenzo-para-dioxin (TCDD)

1, 1,2,2-Tetrachloroethane
Tetrachloroethylene (Perchloroethylene)
p-a,a,a-Tetrachlorotoluene
Tetrafluoroethylene
Tetranitromethane
Thioacetamide
4,4'-Thiodianiline
Thiodicarb
Thiourea
Thorium dioxide
Toluene diisocyanate
ortho-Toluidine
ortho-Toluidine hydrochloride
Toxaphene (Polychlorinated camphenes)
Treosulfan
Trichlormethine (Trimustine hydrochloride)
Trichloroethylene
2,4,6-Trichlorophenol
1,2,3-Trichloropropane
Trimethyl phosphate
2,4,5-Trimethylaniline and its strong acid salts
Triphenyltin hydroxide
Tris(aziridinyl)-para-benzoquinone (Triaziquone)
Tris(l-aziridinyl)phosphine sulfide (Thiotepa)
Tris(2-chloroethyl) phosphate
Tris(2,3-dibromopropyl)phosphate
Trp-P-1 (Tryptophan-P-1)
Trp-P-2 (Tryptophan-P-2)
Trypan blue (commercial grade)
Uracil mustard
Urethane (Ethyl carbamate)
Vinclozolin
Vinyl bromide
Vinyl chloride
4-Vinylcyclohexene
4-Vinyl-l-cyclohexene diepoxide (Vinyl cyclohexenedioxid
Vinyl fluoride
Vinyl trichloride (1,1,2-Trichloroethane)
2,6-Xylidine (2,6-Dimethylaniline)
Zileuton

Chemicals Known to Cause Reproductive Toxicity
Developmental Toxicity

Acetazolamide
Acetohydroxamic acid
Actinomycin D
All-trans retinoic acid
Alprazolam
Altretamine
Amantadine hydrochloride
Amikacin sulfate
Aminoglutethimide
Aminoglycosides
Aminopterin
Amiodarone hydrochloride
Amitraz
Amoxapine
Angiotensin converting enzyme (ACE) inhibitors
Anisindione
Arsenic (inorganic oxides)
Aspirin
Atenolol
Auranofin
Azathioprine
Barbiturates
Beclomethasone dipropionate
Benomyl
Benzene
Benzphetamine hydrochloride
Benzodiazepines
Bischloroethyl nitrosourea (BCNU) (Carmustine)
Bromacil lithium salt
Bromoxynil
Bromoxynil octanoate
Butabarbital sodium
1,4-Butanediol dimethanesulfonate (Busulfan)
Cadmium
Carbamazepine
Carbon disulfide
Carboplatin
Chenodiol
Chinomethionat (Oxythioquinox)
Chlorambucil
Chlorcyclizine hydrochloride

Chlordecone (Kepone)
Chlordiazepoxide
Chlordiazepoxide hydrochloride
1-(2-Chloroethyl)-3-cyclohexyl-l-nitrosourea (CCNU) (Lomustine)
Chlorsulfuron
Cidofovir
Cladribine
Clarithromycin
Clobetasol propionate
Clorniphene citrate
Clorazepate dipotassium
Cocaine
Codeine phosphate
Colchicine
Conjugated estrogens
Cyanazine
Cycloate
Cycloheximide
Cyclophosphamide (anhydrous)
Cyclophosphamide (hydrated)
Cyhexatin
Cytarabine
Dacarbazine
Danazol
Daunorubicin hydrochloride
2,4-D butyric acid
o,p, -DDT
p,p, -DDT
2,4-DP (dichloroprop)
Derneclocycline hydrochloride (internal use)
Diazepam
Diazoxide
Dichlorophene
Dichlorphenamide
Diclofop methyl
Dicumarol
Diethylstilbestrol (DES)
Diflunisal
Dihydroergotamine mesylate
Diltiazem hydrochloride
Dinocap
Dinoseb
Diphenylhydantoin (Phenytoin)
Disodium cyanodithioimidocarbonate

Doxorubicin hydrochloride
Doxycycline (internal use)
Doxycycline calcium (internal use)
Doxycycline hyclate (internal use)
Doxycycline monohydrate (internal use)
Endrin
Ergotamine tartrate
Estropipate
Ethionamide
Ethyl alcohol in alcoholic beverages
Ethyl dipropylthiocarbamate
Ethylene dibromide
Ethylene glycol monoethyl ether
Ethylene glycol monomethyl ether
Ethylene glycol monoethyl ether acetate
Ethylene glycol monomethyl ether acetate
Ethylene thiourea
Etodolac
Etoposide
Etretinate
Fenoxaprop ethyl
Filgrastim
Fluazifop butyl
Flunisolide
Fluorouracil
Fluoxymesterone
Flurazeparn hydrochloride
Flurbiprofen
Flutamide
Fluticasone propionate
Fluvalinate
Ganciclovir sodium
Goserelin acetate
Halazepam
Halobetasol propionate
Haloperidol
Halothane
Heptachlor
Hexachlorobenzene
Histrelin acetate
Hydramethy1non
Hydroxyurea
Idarubicin hydrochloride
Ifosfamide

Iodine-1 31
Isotretinoin
Lead
Leuprolide acetate
Levodopa
Linuron
Lithium carbonate
Lithium citrate
Lorazepam
Lovastatin
Mebendazole
Medroxyprogesterone acetate
Megestrol acetate
Melphalan
Menotropins
Meprobarnate
Mercaptopurine
Mercury and mercury compounds
Methacycline hydrochloride
Metham sodium
Methazole
Methimazole
Methotrexate
Methotrexate sodium
Methyl bromide as a structural fumigant
Methyl chloride
Methyl mercury
Methyltestosterone
Metiram
Midazolam hydrochloride
Minocycline hydrochloride (internal use)
Misoprostol
Mitoxantrone hydrochloride
Myclobutanil
Nabam
Nafarelin acetate
Neomycin sulfate (internal use)
Netilmicin sulfate
Nickel carbonyl
Nifedipine
Nimodipine
Nftrapyrin
Nitrogen mustard (Mechlorethamine)
Nitrogen mustard hydrochloride (Mechlorethamine hydrochloride)

Norethisterone (Norethindrone)
Norethisterone acetate (Norethindrone acetate)
Norethisterone (Norethindrone)/Ethinyl estradiol
Norethisterone (Norethindrone)/Mestranol
Norgestrel
Oxadiazon
Oxazepam
Oxymetholone
Oxytetracycline (internal use)
Oxytetracycline hydrochloride (internal use)
Paclitaxel
Paramethadione
Penicillamine
Pentobarbital sodium
Pentostatin
Phenacemide
Phenprocournon
Pimozide
Pipobroman
Plicamycin
Polybrominated biphenyls
Polychlorinated biphenyls
Potassium dimethyldithiocarbamate
Pravastatin sodium
Prednisolone sodium phosphate
Procarbazine hydrochloride
Propargite
Propylthiouracil
Pyrimethamine
Quazepam
Resmethrin Retinol/retinyl esters, when in daily dosages in excess of 10,000 1 U, or
    3,000 retinol equivalents.
Ribavirin
Rifampin
Secobarbital sodium
Sermorelin acetate
Sodium dimethyldithiocarbamate
Streptomycin sulfate
Streptozocin (streptozotocin)
Sulindac
Tamoxifen citrate
Temazepam
Teniposide
Terbacil

Testosterone cypionate
Testosterone enanthate
2,3,7,8-Tetrachlorodibenzo-para-dioxin (TCDD)
Tetracycline (internal use)
Tetracyclines (internal use)
Tetracycline hydrochloride (internal use)
Thalidomide
Thioguanine
Tobramycin sulfate
Toluene
Triadimefon
Triazolarn
Tributyltin methacrylate
Trientine hydrochloride
Triforine
Trilostane
Trimethadione
Trimetrexate glucuronate
Uracil mustard
Urethane
Urofollitropin
Valproate (Valproic acid)
Vinblastine sulfate
Vinclozolin
Vincristine sulfate
Warfarin
Zileuton

Female Reproductive Toxicity

Aminopterin
Arniodarone hydrochloride
Anabolic steroids
Aspirin
Carbon disulfide
Chlorsulfuron
Cidofovir
Clobetasol propionate
Cocaine
Cyclophosphamide (anhydrous)
Cyclophosphamide (hydrated)
o,p,-DDT
p,p, -DDT
Diflunisal Dinitrotoluene (technical grade)

Ethylene oxide
Etodolac
Flunisolide
Flurbiprofen
Gemfibrozil
Goserelin acetate
Haloperidol
Lead
Leuprolide acetate
Levonorgestrel implants
Nifedipine
Oxydemeton methyl
Paclitaxel
Pimozide
Rifampin
Streptozocin (streptozotocin)
Sulindac
Thiophanate methyl
Triadimefon
Uracil mustard
Zileuton

Male Reproductive Toxicity

Altretamine
Arniodarone hydrochloride
Anabolic steroids
Benomyl
Benzene
Cadmium
Carbon disulfide
Chlorsulfuron
Cidofovir
Colchicine
Cyclohexanol
Cyclophosphamide (anhydrous)
Cyclophosphamide (hydrated)
2,4-D butyric acid
o,p, -DDT
p,p, -DDT
1,2-Dibromo-3-chloropropane (DBCP)
m-Dinitrobenzene
o-Dinitrobenzene
p-Dinitrobenzene

2,4-Dinitrotoluene
2,6-Dinitrotoluene
Dinitrotoluene (technical grade)
Dinoseb
Doxorubicin hydrochloride
Epichlorohydrin
Ethylene dibromide
Ethylene glycol monoethyl ether
Ethylene glycol monomethyl ether
Ethylene glycol monoethyl ether acetate
Ethylene glycol monomethyl ether acetate
Ganciclovir sodium
Gemfibrozil
Goserelin acetate
Hexamethylphosphoramide
Hydramethyinon
Idarubicin hydrochloride
Lead
Leuprolide acetate
Myclobutanil
Nifedipine
Nitrofurantoin
Oxydemeton methyl
Paclitaxel
Quizalofop-ethyl
Ribavirin
Sodium fluoroacetate
Streptozocin (streptozotocin)
Sulfasalazine
Thiophanate methyl
Triadimefon
Uracil mustard

Appendix M

General Drinking Water Monitoring and Warning Requirements

Lead and Copper

Monitoring for lead and copper is required for both Community Water Systems (CWS) and Non-community Water Systems (NCWS). Tap water samples must be collected every 6 months. However, if the system meets the action levels (i.e., Maximum Contaminant Levels, see Appendix A) and maintains optimal corrosion control treatment for two consecutive six-month periods, tap water sampling can be reduced to once a year. For those systems that exceed the action levels, samples of source water must also be collected.

All systems that exceed the lead action level must provide EPA-developed educational materials to their customers within 60 days. A warning must also be added to each water bill and repeated every 12 months as long as the lead exceeds the action level. This warning notice must be given to all daily and weekly newspapers and brochures delivered to specified locations in the service area. In addition, public service announcements must be given every six months to at least five radio and television stations in the service area.

Total Trihalomethane Compounds

CWSs that serve a population of more than 10,000 and add chlorine or bromine disinfectant to the water must monitor for total trihalomethane compounds[1]. If the water source used by the CWS is a surface body of

---

[1]The Stage 2 Disinfection By-Product Rule (May, 2002) will add monitoring for the Haloacetic acids, bromate and chlorite.

water, the CWS must monitor on a quarterly basis, if it is a groundwater source only annual monitoring is needed[2]. However, if a groundwater source exceeds the action level, monitoring must be increased to a quarterly sampling. If the average of four consecutive quarterly samples exceeds the total trihalomethane action level, public notification is required.

Inorganic Chemicals

The Group I inorganic chemicals are barium, cadmium, chromium, mercury and selenium while the Group II inorganic chemicals are antimony, berylium, cyanide, nickel, sulfate and thallium. The Group I and II inorganic chemicals for both CWS and Non-transient Non- community Water System (NTNCWS) must be monitored once every 3 years for groundwater systems and annually for surface water systems. Both systems having at least three previous Group I sampling results below the Maximum Contaminant Level will be granted a waiver and can be then monitored once every nine years. Any supply exceeding the maximum contaminant level must be sampled quarterly for at least one year until the four consecutive samples are below the maximum contaminant level. Public notification is required if the average of the past four quarters exceeds the maximum contaminant level.

All public water systems must monitor for nitrate. CWS and NTNCWS systems that use surface water as their source must monitor quarterly. All public water systems that use groundwater as their source must monitor annually, while NCWS that use surface water as their source must also monitor annually. If the maximum contaminant level is exceeded, the public water supply must notify the public.

All public water systems must monitor for nitrite. Any CWS or NTNCWS that exceed the maximum contaminant level must be sampled quarterly for at least one year. When at least four consecutive samples are less than the maximum contaminant level, monitoring can be reduced to once a year. If the maximum contaminant level is exceeded, the public water supply must notify the public.

CWS and NTNCWS must also monitor for arsenic. Any supply exceeding the maximum contaminant level will be sampled quarterly for at least one year until the average of four consecutive samples are below the maximum contaminant level. Sampling will then resume with one sample

---

[2]Because the formation of trihalomethane compounds results from the reaction of the disinfectant with organic matter in water and groundwater usually has less organic matter than surface water, the frequency of monitoring is reduced for groundwater sources.

every three years for public water systems that use groundwater for their source and annually for surface water sources.

CWS are required to monitor for fluoride. When the average of four original and three repeat samples exceeds the maximum contaminant level, public notification is required.

Volatile Organic Chemicals

CWS and NTNCWS must be monitored for benzene, carbon tetrachloride, o- dichlorobenzene, p-dichlorobenzene, 1,2-dichlorethane, 1,1-dichloroethylene, cis-1,2- dichloroethylene, trans-1,2-dichloroethylene, dichloromethane, 1,2-dichloropropane, ethylbenzene, monochlorobenzene, styrene, toluene, trichloroethylene, 1,1,2-trichloroethane, 1,1,1-trichloroethane, 1,2,4-trichlorobenzene, vinyl chloride and xylenes. If any chemical is detected above the maximum contaminant level, the system must be sampled quarterly. If the average of four consecutive samples is below the maximum contaminant level, sampling can be reduced to once a year for the next three years. If the average of four consecutive quarterly samples exceeds the maximum contaminant level, the system is required to notify the public.

Synthetic Organic Chemicals

CWS and NTNCWS must be monitored for alachlor, aldicarb, aldicarb sulfone, aldicarb sulfoxide, atrazine, carbofuran, chlordane, dalapon, dibromochloropropane, dinoseb, diquat, 2,4- D, endothall, endrin, ethylene dibromide, glyphosate, heptachlor, heptachlor epoxide, lindane, methoxychlor, oxamyl, pentachlorophenol, picloram, PCB, simazine, toxaphene, 2,4,5-TP, benzo(a)pyrene, di-(ethylhexyl)adipate, di-(ethylhexyl)phthalate, hexachlorobenzene, hexachlorocyclopentadiene, and dioxin. If any of these chemicals is detected, the system must be sampled quarterly for at least a year. After one year of quarterly monitoring, the sampling may be reduced to once a year if the results are below the maximum contaminant level. After three years of annual monitoring below the maximum contaminant level, the system must return to four quarterly samples every three years. If the average of four consecutive quarterly samples exceeds the maximum contaminant level, the system is required to notify the public.

Treatment Chemicals

Each public water supply must certify annually that acrylamide and epichlorohydrin do not exceed the maximum contaminant level. No notification is required.

Appendix N

National Drinking Water Contaminant Occurrence Database Data on Primary Water Quality Standards

(Data as of May 18, 2001)

Chemical	Resource(*)	Average (ug/l)	Range (ug/l)		MCL (ug/l)
		Inorganic Compounds			
Antimony	Surface Water(50)	3.059	0.03	— 50.0	6.0
	Ground Water(133)	3.0266	0.2	— 50.0	
Arsenic	Surface Water(98)	165.4088	0.28	— 1000	50.0
	Ground Water(677)	65.4899	0.2	— 1000	
Barium	Surface Water (259)	64.0223	0.2	— 2000	2000
	Ground Water(1196)	145.5923	0.19	— 9200	
Beryllium	Surface Water(42)	36.7646	0.01	— 50	4.0
	Ground Water(123)	27.9408	0.2	— 50	
Cadmium	Surface Water (67)	32.9276	0.01	— 50	5.0
	Ground Water(213)	64.147	0.6	— 2000	
Chromium	Surface Water(106)	242.2496	0.05	— 2000	100.0
	Ground Water(544)	146.7289	0.07	— 14000	

Chemical	Resource(*)	Average (ug/l)	Range (ug/l)	MCL (ug/l)
Copper	Surface Water(60)	269.1479	0.5 – 2580	1300.0
	Ground Water(746)	194.6182	0.2 – 23000	
Cyanide	Surface Water(22)	2844.3333	1 – 10000	200.0
	Ground Water(53)	2194.19	2 – 32700	
Fluoride	Surface Water(316)	609.14	0.77 – 4060	4000
	Ground Water(1627)	720.5208	300 – 1500	
Lead	Surface Water(90)	7.5624	0.10 – 110	15.0
	Ground Water(572)	7.4019	0.1 – 1900	
Mercury	Surface Water(87)	12.8257	0.1 – 25	2.0
	Ground Water(222)	20.7166	0.1 – 3900	
Nitrate	Surface Water(313)	1065.15	10 – 12000	10000
	Ground Water(1943)	2142.55	0.6 – 558000	
Nitrite	Surface Water(47)	67.60	1 – 1770	1000
	Ground Water(231)	62.43	1 – 1800	
Selenium	Surface Water(85)	553.09	0.41 – 14000	50
	Ground Water(320)	143.2883	0.3 – 14000	
Thallium	Surface Water(42)	192.26	0.04 – 10000	2.0
	Ground Water(160)	576.63	0.039 – 43100	
Organic Compounds				
Alachlor (Lasso)	Surface Water(37)	0.7338	0.03 – 2.2	2.0
	Ground Water(9)	1.3458	0.08 – 4.4	
Atrazine	Surface Water(106)	1.8995	0.08 – 42	3.0
	Ground Water(82)	1.1293	0.06 – 12	

Chemical	Resource(*)	Average (ug/l)	Range (ug/l)	MCL (ug/l)
Benzene	Surface Water(42)	3.1768	0.13 — 128	5.0
	Ground Water(157)	104.209	0.098 — 43213	
Benzo(a)pyrene	Surface Water(8)	0.7167	0.02 — 1.0	0.2
	Ground Water(12)	0.3981	0.02 — 0.84	
Carbofuran	Surface Water(6)	0.6417	0.2 — 4.1	40.0
	Ground Water (2)	0.4252	0.2 — 2	
Carbon tetrachloride	Surface Water(92)	1.408	0.1 — 12	5.0
	Ground Water(179)	71.8988	0.13 — 37738	
Chlordane	Surface Water(0)			
	Ground Water(3)	0.43	0.28 — 0.57	2.0
Chlorobenzene	Surface Water(84)	3.8011	0.1 — 38	100.0
	Ground Water(72)	159.5636	0.0009 — 22220	
2,4-D	Surface Water(60)	1.1787	0.1 — 58	70.0
	Ground Water(52)	0.8706	0.083 — 8	
Dalapon	Surface Water(35)	12.1473	0.6 — 68	200.0
	Ground Water(21)	3.9365	0.37 — 25	
DBCP	Surface Water(87)	0.6599	0.00001 — 85	0.2
	Ground Water(146)	0.4261	0.0008 — 100	
O-Dichlorobenzene	Surface Water(41)	2229.81	0.1 — 100100	600.0
	Ground Water(70)	1.1767	0.11 — 21	
p-Dichlorobenzene	Surface Water(50)	5.1068	0.1 — 75	75.0
	Ground Water(94)	0.9326	0.03 — 11	

Chemical	Resource(*)	Average (ug/l)	Range (ug/l)		MCL (ug/l)
1,2-Dichloroethane	Surface Water(44)	6.8174	0.1	– 500	5.0
	Ground Water(125)	2.0671	0.17	– 15	
Cis-1,2-Dichloro-ethylene	Surface Water(53)	408.3739	0.0012	–100100	70.0
	Ground Water(177)	3.7655	0.0008	– 94.8	
1,1-Dichloro-ethylene	Surface Water(42)	7.9406	0.1	– 2300	7.0
	Ground Water(130)	2.457	0.11	– 35.2	
Trans-1,2-Di-chloroethylene	Surface Water(31)	2277.12	0.1	–100100	100.0
	Ground Water(92)	3.7371	0.11	– 61	
Dichloromethane	Surface Water(297)	166.1186	0.0005	–100100	5.0
	Ground Water(1079)	2.576	0.0004	– 620	
1,2-Dichloro-propane	Surface Water(43)	1301.6	0.0005	–100100	5.0
	Ground Water(76)	577.2092	0.0013	–100000	
Di(2-Ethylhexyl)-Adipate	Surface Water(32)	10.2818	0.02	– 180	400.0
	Ground Water(158)	1.1036	0.01	– 132	
Di(2-Ethylhexyl)-Phthalate	Surface Water(15)	2.1824	0.11	– 36	6.0
	Ground Water(181)	0.8267	0.01	– 65	

Chemical	Resource(*)	Average (ug/l)	Range (ug/l)	MCL (ug/l)
Dinoseb	Surface Water(16)	0.5273	0.1 — 4	
	Ground Water(11)	0.4934	0.1 — 2	7.0
2,3,7,8-TCDD-(Dioxin)	Surface Water(0)	0.0		
	Ground Water(0)	0.0		
Diquat	Surface Water(12)	1.8857	0.3 — 24	20.0
	Ground Water(15)	2.7111	0.3 — 41.8	
Endothall	Surface Water(1)	9.0	9.0 — 9.0	100.0
	Ground Water(4)	1142.2	1.9 — 4548	
Endrin	Surface Water(11)	0.0338	0.008 — 0.1	2.0
	Ground Water(10)	0.1487	0.008 — 1.1	
Ethylbenzene	Surface Water(80)	2.6042	0.001 — 100	700.0
	Ground Water(331)	524.6229	0.0008 — 274750	
Ethylene dibromide	Surface Water(88)	1.3187	0.00001 — 28	0.05
	Ground Water(118)	0.2316	0.0001 — 7.14	
Glyphosate	Surface Water (0)			
	Ground Water (2)	9	7 — 11	700
Heptachlor	Surface Water(6)	0.107	0.043 — 0.388	0.4
	Ground Water(9)	0.1764	0.003 — 1.34	
Heptachlor Exoxide	Surface Water(6)	0.0627	0.02 — 0.07	0.2
	Ground Water(12)	0.1095	0.005 — 0.99	
Hexachlorobenzene	Surface Water(5)	0.9859	0.018 — 1.2	1.0
	Ground Water(2)	1.2	1.2 — 1.2	

Chemical	Resource(*)	Average (ug/l)	Range (ug/l)	MCL (ug/l)
Hexachlorocyclo-pentadiene	Surface Water(38)	0.4873	0.051 – 2	50.0
	Ground Water(8)	2.6335	0.184 – 4.4	
Lindane	Surface Water(10)	0.4949	0.005 – 0.8	0.2
	Ground Water(13)	0.3417	0.008 – 0.3417	
Mthoxychlor	Surface Water(9)	0.3738	0.06 – 1	40.0
	Ground Water(6)	0.2729	0.05 – 1	
Oxamyl (Vydate)	Surface Water(3)	1.3	1.3 – 1.3	200
	Ground Water(2)	1.3	1.3 – 1.3	
Polychlorinated Biphenyls	Surface Water(1)	2.9	—	0.5
	Ground Water(6)	0.8485	0.1 – 5.7	
Pentachlorophenol	Surface Water(18)	0.4052	0.04 – 1	1.0
	Ground Water(23)	0.495	0.04 – 1.64	
Picloram	Surface Water(19)	0.7185	0.108 – 2	500
	Ground Water(16)	0.5188	0.11 – 1.1	
Simazine	Surface Water(75)	0.6383	0.07 – 4.89	4.0
	Ground Water(38)	0.4684	0.02 – 2.5	
Styrene	Surface Water(36)	18.4037	0.044 – 660	100
	Ground Water(90)	1.5169	0.03 – 10	
Total Trihalomethanes	Surface Water(196)	40.2774	0.19 – 3295	100
	Ground Water(207)	16.0413	0.08 – 18400	

Chemical	Resource(*)	Average (ug/l)	Range (ug/l)		MCL (ug/l)
2,4,5-TP (Silvex)	Surface Water(16)	0.148	0.04	0.574	50.0
	Ground Water(26)	2.3825	0.01	10.5	
1,2,4-Trichloro-benzene	Surface Water(19)	2.1989	0.5	10	70.0
	Ground Water(68)	1.3614	0.15	21	
1,1,1-Trichloroethane	Surface Water(100)	2.8379	0.1	87	200.0
	Ground Water (355)	8.1193	0.31	2200	
1,1,2-Trichloro-ethane	Surface Water(55)	1.9938	0.0005	85	5.0
	Ground Water(70)	1.3514	0.15	11.7	
Tetrachloroethylene	Surface Water(114)	5.6896	0.0065	2900	5
	Ground Water(544)	5.6103	0.0004	1400	
Toluene	Surface Water(153)	2.957	0.0005	31.3	1000
	Ground Water (473)	818.33	0.0005	574750	
Trichloroethylene	Surface Water(93)	9.3806	0.1	590	5.0
	Ground Water(313)	7.4014	0.01	3668.9	
Vinyl Chloride	Surface Water(26)	1.0993	0.07	7.3	2.0
	Ground Water(57)	1.3955	0.13	12.7	
Xylenes	Surface Water(103)	5.0308	0.001	110	10000
	Ground Water(405)	1639.84	0.053	1334750	

* = Number of Community Water Systems that detected a specific chemical in the drinking water distributed to their consumers.

Appendix O

National Drinking Water Contaminant Occurrence Database Data on Unregulated Compounds

Chemical Data Summary—Group I

Chemical	Resource (*)	Average (ug/l)	Range (ug/l)	
Cyanazine (Bladex)	Surface Water(72)	1.8968	0.31	12
	Ground Water (9)	2.9477	0.449	12
Methyl-tert-butyl-ether	Surface Water(19)	2.0455	0.5	8
	Ground Water(103)	16.1573	0.4	87
Toxaphene	Surface Water(3)	0.7	0.7	0.7
	Ground Water(2)	0.625	0.1	0.7
Trifluralin	Surface Water(7)	0.0771	0.032	0.2
	Ground Water(1)	0.056	0.056	0.056

Chemical Data Summary—Group II

Chemical	Resource (*)	Average (ug/l)	Range (ug/l)	
Aldicarb	Surface Water(29)	3917.8898	0.5	–231000
	Ground Water(8)	848.9462	0.5	– 10000
Aldicarb Sulfone	Surface Water(50)	2016.1246	0.3	–231000
	Ground Water(16)	505.0142	0.0001	– 10000
Aldicarb Sulfoxide	Surface Water(52)	2164.2112	0.0088	–231000
	Ground Water(13)	552.2183	0.0019	– 10000
Aldrin	Surface Water(16)	1.2834	0.07	– 4.7
	Ground Water(19)	0.1976	0.07	– 0.84
Butachlor	Surface Water(8)	11.3093	0.04	– 24
	Ground Water(6)	5.2226	0.04	– 17
Carbaryl	Surface Water(32)	1.8448	0.18	– 4
	Ground Water (8)	734.5587	0.18	– 10000
Dicamba	Surface Water(71)	5.0339	0.0007	– 500
	Ground Water(39)	29.546	0.026	– 1600
1,3-Dichloropropene	Surface Water(35)	2328.6805	0.0005	–100100
	Ground Water(86)	0.827	0.08	– 39
Trans-1,3-Dichloropropene	Surface Water(6)	0.5182	0.2	– 0.7
	Ground Water(7)	0.675	0.3	– 1.7
Dieldrin	Surface Water(12)	2.1903	0.004	– 4.4
	Ground Water(12)	0.2749	0.0001	– 1.65

Chemical	Resource (*)	Average (ug/l)	Range (ug/l)
3-Hydroxycarbofuran	Surface Water(44)	2.7073	0.0015 — 66.3
	Ground Water(13)	577.8382	0.0015 — 10000
Methomyl	Surface Water(29)	1.4984	0.29 — 3
	Ground Water(14)	525.0304	0.0001 — 10000
Metolachlor	Surface Water(234)	1.5316	0.00001 — 130
	Ground Water(53)	82.9208	0.0007 - 10000
Metribuzin	Surface Water(36)	4.7731	0.1 — 230
	Ground Water(11)	358.7057	0.0001 — 10000
Propachlor	Surface Water(40)	1.1259	0.0001 — 3
	Ground Water(5)	910.4548	0.0025 2 10000

Chemical Data Summary - Group III

Chemical	Resource (*)	Average (ug/l)	Range (ug/l)
Bromobenzene	Surface Water(31)	6.1237	0.0034 — 43.2
	Ground Water (80)	256.1072	0.0005 — 22220
Bromochloromethane	Surface Water(64)	4.4931	0.0005 — 42.1
	Ground Water(131)	3.6694	0.0001 — 210
Bromomethane	Surface Water(74)	4.9135	0.0005 — 49.9
	Ground Water(246)	73.1414	0.0001 — 22220
N-Butylbenzene	Surface Water(26)	1.8158	0.0001 — 10
	Ground Water(76)	1.8354	0.0005 — 84
Sec-Butylbenzene	Surface Water(16)	4.4421	0.0005 — 22
	Ground Water(59)	223.2537	0.0005 — 15590

Chemical	Resource (*)	Average (ug/l)	Range (ug/l)
Tert-Butylbenzene	Surface Water(15)	2.4667	0.0002 — 10
	Ground Water(66)	0.8206	0.0005 — 9
O-Chlorotoluene	Surface Water(46)	15.5815	0.0005 — 239
	Ground Water(80)	240.439	0.0005 — 22220
P-Chlorotoluene	Surface Water(43)	16.1985	0.0005 — 239
	Ground Water(65)	314.4116	0.0002 — 22220
Dibromomethane	Surface Water(99)	714.805	0.0005 —100100
	Ground Water(141)	1.595	0.0005 — 21
M-Dichlorobenzene	Surface Water(41)	2090.1363	0.0005 —100100
	Ground Water (88)	0.7825	0.0005 — 21
1,1-Dichloroethane	Surface Water(59)	722.7888	0.0004 —100100
	Ground Water(285)	2.5805	0.0001 — 500
Dichlorodifluoro-methane	Surface Water(50)	3.6793	0.0005 — 88
	Ground Water(211)	8.0765	0.0001 — 556.9
1,3-Dichloropropane	Surface Water(20)	4767.693	0.0005 —100100
	Ground Water(65)	0.4645	0.0005 — 3.7
2,2-Dichloropropane	Surface Water(21)	1.2459	0.0005 — 6
	Ground Water(63)	1.4564	0.0005 — 61
1,1-Dichloropropene	Surface Water(25)	3851.4901	0.0001 —100100
	Ground Water(59)	0.4812	0.0005 — 5
Hexachlorobutadiene	Surface Water(23)	1.7616	0.0005 — 10
	Ground Water(14)	0.4929	0.0005 — 8

Chemical	Resource (*)	Average (ug/l)	Range (ug/l)
Isopropylbenzene	Surface Water(18)	0.5178	0.0005 – 2.6
	Ground Water(102)	197.448	0.0005 – 28463
P-Isopropyltoluene	Surface Water(17)	0.5496	0.0005 – 5
	Ground Water(68)	466.3746	0.0005 – 35325
Naphthalene	Surface Water(65)	2.1892	0.0001 – 55
	Ground Water(217)	774.3238	0.0001 –243700
N-Propylbenzene	Surface Water(22)	1.9701	0.0001 – 21
	Ground Water (83)	1952.4286	0.0005 –206763
1,1,1,2-Tetrachloro-ethane	Surface Water(31)	0.6672	0.0005 – 6.1
	Ground Water(101)	61.6757	0.0005 – 2700
1,1,2,2-Tetrachloro-ethane	Surface Water(30)	0.5651	0.0003 – 2
	Ground Water(96)	0.8696	0.0005 – 11
1,2,3-Trichlorobenzene	Surface Water(35)	1.4514	0.0005 – 10
	Ground Water(102)	0.5478	0.0004 – 7
Trichlorofluoromethane	Surface Water(107)	1.8953	0.0001 – 45
	Ground Water(291)	8.5342	0.0001 – 1444
1,2,3-Trichloropropane	Surface Water(31)	2.1565	0.0005 – 21
	Ground Water(84)	31.1628	0.0001 – 3000
1,2,4-Trimethylbenzene	Surface Water(61)	1.3424	0.0005 – 16.8
	Ground Water (202)	3390.2383	0.0001 –978250

Chemical	Resource (*)	Average (ug/l)	Range (ug/l)
1,3,5-Trimethyl-benzene	Surface Water(35)	1.5561	0.0002 – 10.9
	Ground Water(137)	1523.8481	0.0005 –288750

* = Number of Community Water Systems that detected a specific chemical in the drinking water distributed to their consumers.

Group I: Approximately 2,500 Community Water Systems conducted analyses for the listed chemicals.

Group II: Approximately 10,000 Community Water Systems conducted analyses for the listed chemicals.

Group III: Approximately 20,000 Community Water Systems conducted analyses for the listed chemicals.

Data collected as of May 18, 2001

Appendix P

Components of
In-Home Treatment Systems

As discussed previously, a combination of currently available treatment systems can effectively reduce most pollutants to undetectable levels. However, additional research is still necessary to pinpoint the actual removal efficiencies of treatment units when dealing with certain regulated and unregulated chemicals. Even with this uncertainty, a combination of technologies can still be used to produce water that contains a minimum concentration of chemical pollutants when the system is properly designed, maintained and the final product stored in a glass lined tank (or a comparable tank that does not leach chemicals into the contained water).

Both reverse osmosis (RO) and granulated activated carbon (GAC) units are capable of removing anywhere from 95 to 99 percent of the pollutants that pass through them. When two different units are used in series, the removal efficiencies can be multiplied to determine the total percent pollutant removal. For example, if a water resource with 10 parts-per-billion arsenic (i.e., the future arsenic standard in drinking water) was treated using RO, the water produced would contain only 0.5 parts-per-billion arsenic. If this water was further treated using an activated alumina[1] unit, the final effluent would only contain 0.25 parts-per-billion, or 25 parts-per-trillion arsenic. If unit efficiencies of 99 percent and 75 percent are assumed, the resulting concentration would approach 3 parts-per-trillion. Although this is not "zero" pollution, it is very close. While not all chemicals are removed at these high efficiencies, many chemicals are readily removed by reverse osmosis (as shown in Table 5, Chapter 3) or by carbon

[1]Torrens, Kevin D., "Evaluating Arsenic Removal Technologies," *Pollution Engineering* (July 1999).

filters (as shown in Table 4, Chapter 3) over a range of removal efficiencies[2]. Thus, two or more treatment units operating in series will often provide the best possible overall removal efficiencies.

The cost of an in-home water treatment system is more a function of water usage than chemical constituents present in the water. The greater the water demand the greater the system capital and maintenance costs. Therefore, if cost is a major consideration, a home owner may have to settle for a point-of-use system (e.g., at the kitchen sink) as opposed to the more desirable point-of-entry system.

Point-of-Use Systems

Under-the-sink point-of use water treatment units are widely available through many retail and internet retailers. Retailers offer a range of treatment options from simple sediment filters to single and combined RO and GAC treatment units. Occasionally these units are sold in some combination but more usually as stand-alone units. As discussed in Chapter 3, a good point-of-use system[3] at the kitchen sink should include a pre-filter to remove suspended solids and chlorine, a canister system containing at least an OR unit followed by at least one GAC filter unit, a 4–5 gallon storage tank and an ultraviolet light treatment unit between the storage tank and faucet. Water treatment units in this standard configuration generally range in cost from $500 to $800. If the treated water pressure is low, a booster pump may be required at an additional $200. This price is for the equipment only and does not include installation. Depending upon where the unit is located (i.e., under the sink, utility closet, basement or garage), installation may cost from $150 to $500. Therefore, the consumer can expect to spend up to approximately $1,500 for an installed point-of-use system.

In addition to the installed cost of the treatment, filters, membranes and ultraviolet lamps must be periodically replaced. Annual costs of sediment and carbon filter replacement range from approximately $75 to $100, an OR membrane can range from $45 to $75, and an ultraviolet lamp from $50 to $60. Thus, annual maintenance costs should range from $170 to $235. In addition to those costs, a homeowner may decide to pay more to upgrade the system:

---

[2]Mindful of the fact that there is such a contaminant as organic arsenic, carbon treatment is effective at removing this form of arsenic at efficiencies approaching 75 percent.

[3]This example assumes that the source water is from a community water supply that has been disinfected, contains low levels of iron and manganese (less than 0.3 ppm) and does not exceed a general hardness of 140 to 210 ppm.

- In order to increase the performance of the OR membrane, the home-owner may decide to install a point-of-entry water softener and iron removal unit.
- If the water source is untreated (i.e., a private surface water or ground-water source), pre- filtration for sediment control as well as disinfection for bacterial control might be required prior to OR or GAC water treatment.

In-home water treatment systems are neither cheap nor maintenance-free and they definitely require routine maintenance. Without maintenance, both RO and GAC units will cease removing pollutants and, in fact, can pose a biological hazard. Therefore, it is critical that the consumer follow all the manufacturer and/or retailer guidelines and recommendations on both installation and maintenance of the system. A simple alternative that deals with this problem is to retain a water treatment service company to both monitor and maintain the system.

Point-of-Entry Systems

Residential and commercial point-of-entry systems are available today but usually require a technician for assembly, installation and testing. Because of the fairly limited demand for minimum-pollution point-of-entry systems, there are only a limited number of suppliers. When coupled with the cost of "custom installation," such systems are costly even without any required pre-conditioning of water. If the source is from a community water supply that has been disinfected, is low in iron and manganese and is generally soft, no pre-conditioning is necessary. Otherwise, pre-conditioning systems will need to be installed. Such systems may include sediment removal followed by disinfection, water softening and iron/manganese removal as mentioned above.

Point-of-entry systems usually use just a single technology such as reverse osmosis or carbon filtration (with or without ultraviolet light). However, custom systems can be ordered that include several treatment components. Not all point-of-entry systems require custom design but they do require custom installation. The basic component of any point-of-entry system is the RO treatment unit. An RO package, containing the following components, can be obtained from many retailers today:

- a five-micron sediment pre-filter,
- a carbon block pre-filter to remove chlorine,
- a 500-gallon-per-day RO unit,

- a 300-gallon storage tank,
- an automatic low pressure shutoff value,
- a repressurization system to provide treated water at up to 8 gallons per minute, and
- a digital total dissolved solids meter to monitor system performance.

The cost for this package ranges from \$4,900 to \$5,500[4]. It is also strongly recommended that a GAC filter be added following the RO unit.

Typical activated carbon filters produce water at rates from five to ten gallons per minute and at costs ranging from approximately \$550 to \$1,000. The effective operational life (before cartridge replacement) is typically one to two years based on the volume of water treated and the amount of chemicals removed. Both carbon filters and RO units of this size automatically backwash[5] on a regular basis so as to maintain system performance.

The treated water is usually stored in a fiberglass tank. If possible, water should be stored in a glass lined tank. An ultraviolet system should be placed between the tank outlet and the water inlet to the house to insure that the stored water is disinfected. A seven-gallon per-minute ultraviolet system will range in cost from approximately \$400 to \$750 dollars. Thus, the overall cost of a delivered point-of-entry system could range from approximately \$5,850 to \$7,000[6].

Tom Martin of the Good Water Company, reported that in 2001 they sold 10 to 15 systems of the type described above. If sales were to increase 10- to 100-fold annually, he estimated a corresponding price decrease between 10% to 20%. Needless to say, if this approach to achieving pollution free water were to result in annual sales closer to several hundred thousand or more systems annually, the unit price might well be a third to one-half of today's prices.

As would be expected, these systems have maintenance costs that are greater than point- of-use systems, though not much greater. Based on the experience of the Good Water Company, the annual maintenance costs for

---

[4]All referenced costs are based on 2001 prices.

[5] The residue from home water systems, specifically from sediment traps and reverse osmosis units, goes into the sanitary system and since that which is backwashed is simply an accumulation of what was in the water originally, the actual amount discharged with the wash water is identical to what would have gone down the drain anyway. The fraction that is caught and removed by GAC is actually removed as part of the carbon regeneration process. Thus, in the final analysis, the waste from a home system should not adversely affect the sewage treatment plant to which it is sent.

[6]Given the variables involved with site-specific setup, piping and testing, it is not possible to estimate the ultimate installed cost.

replacing filters, membranes and ultraviolet lamps have averaged approximately $350. Once again, it is critical that the consumer follow all the manufacturer and/or retailer guidelines and recommendations for installation and maintenance. With point-of-entry systems, it is recommended that a water treatment service be hired to monitor and maintain these system.

Index

235

Environmental Chemistry and Human Health Institute

The goals of this non-profit Institute are to further the understanding of the relationship between chemical compounds within the man-made environment and human health and to recommend science policy that will advance the protection of the public health. These goals are accomplished by conducting and/or sponsoring research activities that (1) explore the distribution of chemicals in food, water, air and land, (2) investigate the total chemical exposure (i.e., food, water and air) of the most sensitive members of society (e.g., children, pregnant women, the aged, individuals with immune deficiencies, individuals under stress), (3) evaluate methods of limiting chemical exposures, (4) carry out policy and economic analyses that will support the reduction of chemical exposures, and (5) develop materials suitable for educational purposes.

Donations to the Institute can be arranged by contacting Dr. Patrick Sullivan at the Environmental Chemistry and Human Health Institute, 60 E. Third Avenue, Suite 385, San Mateo, California, 94401, (650) 347-1466 or at http://www.psfmaenv@pacbell.net.

Practical Environmental Forensics:
Process and Case Histories

By Patrick J. Sullivan, Franklin J. Agardy, and Richard K. Traub

Environmental forensics is the study, analysis, and evaluation of environmental issues in a legal dispute. Involving much more than chemical analyses or modeling, it is a complex process that requires establishing the truth for each and every technical and often competing element at issue in an environmental contamination case. *Practical Environmental Forensics: Process and Case Histories* (Wiley; Hardcover; 0-471-35398-1), is the first book to offer a comprehensive look at this exciting new interdisciplinary approach to the environmental litigation process.

The primary goal of *Practical Environmental Forensics* is not only to help engineering and scientific professionals understand the forensic process as it pertains to environmental problems, but also to provide insightful information to potential experts, including those involved in construction, intellectual property, patents, medicine, product liability, and personal injury litigation. A secondary goal is to provide technical professionals and attorneys with an in-depth look at both the environmental law and the scientific issues commonly encountered in environmental cases.

Practical Environmental Forensics can be ordered at www.wiley.com or other Internet bookstores such as amazon.com and barnesandnoble.com.

ISBN 1553696166